Antoine Lavo

ANTOINE LAVOISIER

SCIENCE, ADMINISTRATION, AND REVOLUTION

ARTHUR DONOVAN

Published by the Press Syndicate of the University of Cambridge
The Pitt Building, Trumpington Street, Cambridge CB2 1RP
40 West 20th Street, New York, NY 10011-4211, USA
10 Stamford Road, Oakleigh, Melbourne 3166, Australia

First published 1993 by Blackwell Publishers, Oxford
Reissued by Cambridge University Press 1996

British Library Cataloguing in Publication Data available

Library of Congress Cataloging-in-Publication Data available

ISBN 0 521 56218 X hardback
ISBN 0 521 56672 X paperback

Transferred to digital printing 1999

Contents

CONTENTS

Illustrations

Illustrations 1 and 2 were originally published in Lavoisier, *Oeuvres* (vol. 5, plate I; vol. 3, plate IX); numbers 7 and 8 were first published in Grimaux's biography. The author wishes to acknowledge with thanks the provision of prints and permission to publish them from the following institutions: 1, 5, and 6, Division of Rare and Manuscript Collections, Carl A. Kroch Library, Cornell University; 2, 7, and 8, Special Collections, Van Pelt-Dietrich Library, University of Pennsylvania; 3, Département des Estampes, Bibliothèque Nationale, Paris; 4, Metropolitan Museum of Art, New York.

General Editor's Preface

Our society depends upon science, and yet to many of us what scientists do is a mystery. The sciences are not just collections of facts but are ordered by theory, and this is where Einstein's famous phrase about science being a free creation of the human mind comes in. Science is a fully human activity; the personalities of those who practice it are important in its progress and often interesting to us. Looking at the lives of scientists is a way of bringing science to life.

Maybe what most people know about Lavoisier was that he was beheaded during the Reign of Terror in France. Before reaching the scaffold he had transformed the science of chemistry, with his theory that combustion involved combination with *oxygen* rather than emission of *phlogiston*; with his emphasis upon *weighing*; and with his presiding over the introduction of a new *language* for the science. The book outlining his new chemistry was published in 1789, the year in which the Bastille was stormed and the French Revolution began; and he saw his own work as a chemical revolution in which inherited but obsolete and ineffective ways were replaced by a new outlook, new theory, and new methods. Eighty years on Adolph Wurtz could say "Chemistry is a French science; its founder was Lavoisier of immortal fame." This tells us something about French nationalism, but there is truth behind it. Lavoisier is of enormous importance in the history of science, and modern ideas and arguments about how science progresses or changes, in evolutionary or revolutionary ways, involve using his work as one of the prime examples of a dramatic transformation. His chemical revolution has attracted a great deal of scholarly interest and attention in recent years; this book is particularly apt because 1994 marks the bicentenary of his execution.

Lavoisier did not spend very much of his time in the laboratory. People sometimes refer to men of science of his time as "amateurs"; if that means incompetent, it certainly does not apply to him, but if it simply means that science was not his job, then it was true of him and most of his contemporaries. Arthur Donovan, drawing upon recent scholarship and his own extensive research, portrays the man as well as the scientist. We see an ambitious person getting on in the pre-revolutionary years, trying to modernize the administration of government, especially the tax system, marrying well, and becoming a very important figure.

Then comes the chemistry: not really a French science, but with its most important work being done in Germany, Sweden, and Britain, where the isolation of different gases was the most striking phenomenon. Lavoisier's experiments were beautifully planned and carried on in splendid apparatus, and he first saw clearly how the new discoveries could be organized into a science if one adopted a different theory of burning. He was very conscious that it is not enough to hit upon truth; it must be propagated. Through the Academy of Sciences and through his writings he built up a team of disciples, though not everybody cared to be seen as part of the chorus supporting his solo. By the time of his death his new chemistry was coming to prevail all over Europe.

He and others like him had been supporters of the Revolution, believing that France needed energetic modernizing, but his long and intimate association with the tax system ultimately proved his undoing. In Donovan's splendid study Lavoisier's history and that of France are intimately linked. He was a man of his time and his particular culture, the administrative nobility of France. He approached chemistry in the spirit of an accountant, with books to balance and sense to be made of reactions as of transactions; and in the Academy of Sciences he had backing and contacts with those in other branches of science (such as mathematics and experimental physics) which could not but be stimulating. He was a man perhaps more admired than loved. Donovan gives us a good idea of what he was like and of how his mind worked, and makes him accessible to us. I warmly commend the book.

David Knight
University of Durham

Preface

The bicentennial of Lavoisier's death, commemorated in 1994, occasioned the appearance of several significant scholarly works. One of these is Jean-Pierre Poirier's *Antoine Laurent de Lavoisier, 1743–1794* (Paris: Pygmalion, 1993), a comprehensive biography based on extensive archival research. Poirier pays particular attention to Lavoisier's economic ideas and financial affairs; he provides, for instance, an informative account of the notes Lavoisier made in 1772 while reading David Hume's *Essays on Commerce.* Poirier also quotes extensively from letters that throw new light on the lengthy affair between Pierre-Samuel Dupont and Mme Lavoisier. While Poirier's portrait of Lavoisier differs in no essential way from the one presented in the book that follows, those seeking more detail about Lavoisier's many administrative activities should turn to this important study.

Bernadette Bensaude-Vincent's *Lavoisier, Mémoires d'une révolution* (Paris: Flammarion, 1993) complements Poirier's biography in several ways. While thoroughly grounded in the published literature and archival research, her study focuses on the context of ideas and beliefs in which Lavoisier worked rather than the details of his life. Being a philosopher of science, Bensaude-Vincent addresses questions such as what it means to talk of a revolution in chemistry and to what extent Lavoisier employed a distinctive method. She also brings her story into the present by analyzing nineteenth- and twentieth-century images of Lavoisier as a hero of science and progress.

Bensaude-Vincent has also, with Ferdinando Abbri, edited a valuable collection of essays presented at a 1994 conference sponsored by the European Science Foundation: *Lavoisier in European Context – Negotiating a New Language* (Canton, Mass.: Science History Publications/USA, 1995). This collection should be read in conjunction with Marco Beretta's comprehensive and illuminating *The Enlightenment of Matter – The Definition of Chemistry from Agricola to Lavoisier* (Canton, Mass.: Science History Publications/USA, 1993).

Volume V of Lavoisier's *Correspondence*, covering the years 1787–88, was also published in 1993. Edited by Michelle Goupil, this volume makes available documents that add arresting detail to what is known about Lavoisier and his circle during these eventful years. It contains, for instance, a letter that Hassenfratz, a chemist and friend of the Lavoisiers, wrote to Mme Lavoisier on 20 February 1788. She had asked him for suggestions concerning the portrait that the artist J.-L. David would be painting of the Lavoisiers. Hassenfratz's response adds considerably to what little is known about the composition of this most famous of all scientific portraits. Mme Lavoisier's own artistic training is the subject of another bicentennial publication (Madeleine Pinault Sørensen, "Madame Lavoisier, dessinatrice et peintre," *La revue*, published by the Musée des arts et métiers, no. 6, March 1994); this article provides the first documentary proof that Mme Lavoisier had in fact studied drawing in David's studio.

Historic commemorations serve many purposes; happily the bicentennial of Lavoisier's death stimulated the production of several works of enduring value. While much has been written about the entwined topics of Lavoisier, the eighteenth-century chemical revolution, and science in the French Revolution, these subjects are far from exhausted. New evidence and new perspectives continue to appear and re-examinations of Lavoisier and his time continue to reward those drawn to these topics.

Arthur Donovan
Kings Point
August 1995

Acknowledgements

Anyone interested in the origins and nature of Lavoisier's achievements must contend with a vast and highly diverse body of scholarship. My debt to this literature is profound and I acknowledge with gratitude that the authors cited, several of whom I count as mentors, colleagues, and friends, have provided many of the facts and interpretations that inform this biography. When obliged to compress the discussion of certain topics, I have indicated in the notes where more extensive treatment can be found.

Those familiar with the literature on eighteenth-century science will recognize that, at a more detailed level, the story I tell draws from the work of, among others, Henry Guerlac, Carleton Perrin, Maurice Daumas, Charles Gillispie, William Smeaton, Keith Baker, Roger Hahn, Maurice Crosland, Robert Darnton, Frederic Holmes, John Heilbron, Jerry Gough, and Rhoda Rappaport. While I have usually followed happily where they and others have led, I have no doubt on occasion misunderstood or otherwise mangled particular topics and texts. In the best of all possible worlds I could cheerfully attribute these slips to innocent misperception. In the less sunny world we actually inhabit my personal responsibility for the errors that remain cannot be so easily evaded.

And finally, at the level of fine detail, I acknowledge with heartfelt thanks the invaluable contributions of those friends and colleagues who read and commented on the first draft of this work. This list includes William Smeaton, who saved me from innumerable errors of detail, Frederic Holmes, who has repeatedly and correctly reminded me that my treatment of Lavoisier's science is far from complete, Marco Beretta, Lorraine Daston, Rachel Laudan, Seymour Mauskopf, and two colleagues at the US Merchant Marine Academy, Jacques Szaluta and Albert Stwertka. Their encouragement and criticism have improved this biography immeasurably.

Preparation of this book was underwritten in part by grants from the Research Division of the National Endowment for the Humanities and from the Program for the History and Philosophy of Science and Technology of the National Science Foundation; I am truly grateful for this support. I also wish to thank Dean Warren Mazek and the members of the Department of Humanities of the United States Merchant Marine Academy for freeing me from many of my normal duties while I was completing this study. Thanks are also due to Professor Robert Darnton, Princeton University, to Dr Frances Kohler, University of Pennsylvania, and to Margaret Rogers, Cornell University, for providing prints of the illustrations. Other individuals too numerous to mention provided invaluable assistance in libraries and archives and commented helpfully on preliminary papers and drafts. I think of their willingness to share their time, knowledge, and enthusiasm whenever I hear mention of "the community of scholars."

If this biography aspired to being a definitive account of Lavoisier's life and work, the text and notes would bristle with critical assessments of the findings and interpretations of specialists. Conducting a secondary campaign of this sort would have required a much more extensive scholarly apparatus than could be included in this book. My decision to emphasize the flow of the narrative should therefore not be taken as an indication that there is a high degree of consensus among scholars on the many issues addressed. Those wishing to sample the disputatious delights available in less austere studies of Lavoisier are referred to the works listed in the notes. Two convenient points of entry are Guerlac, *Lavoisier*, and Donovan, "Lavoisier and the Origins of Modern Chemistry."

Readers should also know there are masses of Lavoisier manuscripts that have not yet been adequately edited or interpreted. The monumental nineteenth-century edition of Lavoisier's works, although indispensable, omits many important documents and contains significant errors. The publication of his correspondence was being brought to completion under the able editorship of Michelle Goupil prior to her untimely death. The manuscript records of Lavoisier's scientific career deposited in the archives of the Academy of Sciences in Paris are at long last being adequately preserved and catalogued. I have perused these papers and those in the Cornell University Library, as have many others, and I have learned much from monographs based on detailed examinations of these records. As Lavoisier specialists know, these repositories contain many documents that deserve further examination. Other archives in France and elsewhere also contain additional papers of direct relevance to various aspects of Lavoisier's public life. Whether close study of these diverse manuscripts will significantly alter our current picture of Lavoisier remains to be seen; that such study would place our understanding of the man and his achievement on a firmer footing is beyond question.

All translations, unless otherwise indicated, are my own.

Arthur Donovan
Kings Point
March 1993

Introduction

In the second half of the eighteenth century Antoine Lavoisier (1743–94) developed a novel set of theories that transformed the science of chemistry. Thanks to this achievement he has, along with Copernicus, Galileo, Newton, Darwin, and Einstein, long been recognized as a leader of one of the classic scientific revolutions in modern history. His contributions to the chemical revolution of the eighteenth century, that century's only canonical scientific revolution, have earned him the title of the founder of modern chemistry and a place in the pantheon of scientific immortals.

Lavoisier was guillotined during the Terror in 1794 and hence is also remembered as one of the French Revolution's most famous victims. What his execution tells us about the connections between scientific and political progress is still a matter of considerable debate. When thinking about the relationship between his science and his political fate, it is important to remember that, in addition to being a great scientist and a political victim, Lavoisier was a prominent administrator during the final decades of the old regime. He was deeply involved in governmental affairs as a tax official and financier, as a director of the Gunpowder Administration, as a spokesman for the Academy of Sciences, and as a loyal citizen who at the outset optimistically welcomed the political revolution that in the end took his life.

This then is the story of a great scientist who played a leading role in a number of public forums in Paris, the cultural capital of Europe, during a period of extraordinary intellectual and political ferment. The account that follows unfolds chronologically and ranges widely. The narrative is organized around several themes that were of fundamental importance to Lavoisier throughout his life. Once these themes have been identified, it becomes evident that many of his achievements as a scientist were informed by beliefs that shaped other aspects of his career as well. While my goal has been to provide an integrated account of his life, I have anchored this account in specifics by describing particular historical events, by reading selected texts closely, and by tracing out certain cognitive and institutional continuities. If this book succeeds, it will render the public career of this complex individual more intelligible, his scientific achievement more accessible, and the meaning of his life and death more comprehensible.

Readers more impressed by the costs of science than its benefits and those who are still cheered by the overthrow of the old regime may find this account of Lavoisier insufficiently censorious. Biographers are often seduced by their subjects, and the arts and consequences of seduction can be undeniably charming. I freely confess that after years of living with Lavoisier I have grown to admire him, yet in writing this account of his life I have not sought to bathe him in a theatrical aura of goodness and grandeur. I set out to examine his life and career evenhandedly and in the end I found nothing to condemn in his conduct as an individual. Not being inclined to pass judgment on entire epochs or classes, I have told the story of his life as I found it. Lavoisier was most certainly privileged and no doubt he was frequently haughty and impatient, but the biographer's emotional responses to his subject's social position and manners are not matters of great importance. When weighed in the scales of his own age or as a figure in the history of science, Lavoisier commands respect and admiration. If those who think otherwise know of evidence that justifies a more critical appraisal of him, I look forward to seeing it.

This book is not a case study written to support a particular theory of scientific change, nor does it attempt to provide either an exhaustive account of Lavoisier's life or a comprehensive assessment of his contributions to the development of modern chemistry. It is, quite simply, an historical biography written for those who know of Lavoisier as a scientist and wish to get a better sense of the man and his times, and for those who know of Lavoisier as a prominent figure in the French Enlightenment and wish to know more about his science. Perhaps a study that emphasizes how a great scientist at the beginning of the modern age constructed and managed a wide-ranging and influential public career can throw some light on our understanding of the cultural role of science two centuries after his death. There is, after all, nothing wrong with looking to history for lessons that can be applied to the present.

Lavoisier was frequently involved in controversies and any account of his life must therefore grapple with such contentious topics as the status and uses of science in the Enlightenment, the treatment of scientists during the political Revolution, and the social circumstances that encouraged the rise of modern science. My intention throughout has been to build upon the best-informed and most illuminating scholarly treatments of these issues. Recent studies of the Enlightenment and the French Revolution have demonstrated there is still much to be learned about the era and culture in which Lavoisier made his career; the history of modern science is being revised in an equally dramatic manner. If our accounts of Lavoisier, of eighteenth-century science, and of the interaction of science and politics are to remain current, they must incorporate the spirit and substance of these recent historiographic advances.

The narrative that follows unavoidably makes use of a number of familiar but potentially misleading concepts. It would be tiresome to fuss over each of them individually, but a general caution on the dangers of anachronism is in order. Terms such as science, experiment, chemistry, culture, politics, and revolution are all in active use today and carry connotations that most readers routinely treat as unproblematic. The same is true of such technical terms as elements, atoms, and molecules, all of which were employed in eighteenth-century scientific discourse. I have used these terms as convenient bridges between the present and the past. It should not be assumed, however, that the reader's late twentieth-century understanding of chemistry, revolution, or molecule accurately represents the meanings attributed to these terms in the eighteenth century. Reading, happily, imposes interpretive demands on the reader. I have done my best to make the style and argument of this book transparent, yet the reader should remember that in the final analysis the meaning of certain terms and the implications of certain events may not be exactly what they seem on first encounter.

The vocabulary of politics raises the most daunting problems in this regard. The French Revolution was an event of such epochal importance in modern political history that an act of considerable imagination is required to see behind it. We must pierce this veil if we wish to understand how Lavoisier's science arose from and fitted into the culture of pre-revolutionary France. To acquire some sense of the world Lavoisier inhabited, one must try to conceive of politics and culture as he encountered them as a young man. Only then can one appreciate the ways in which he orchestrated the interests, ideas, and institutions at his disposal while constructing his career as a scientist and administrator.

French politics in the eighteenth century can be seen as a public drama with a plot governed by three competing conceptions of political legitimacy.[1] The first of these political visions, described in more detail in parts I and II below, emphasized enlightened absolutism and administrative centralization. By the 1760s, when Lavoisier entered public life, this Bourbon strategy of governance had been pursued for over a century and had brought great power and glory to the French nation.

Proponents of political absolutism considered the king the sole representative of the nation; this was the fundamental sense in which his power was thought to be absolute. Absolutism was a strategy designed to banish forever the kind of factional politics that had reduced French public life to a shambles in the middle of the seventeenth century. The aura of absolutism made it appear that this public life had been thoroughly depoliticized. In fact, however, the king's power was severely limited and he was obliged by the realities of finance, communication, and administration to treat with deference the rights and privileges of France's various social orders and corporations. While the Bourbon kings did not openly tolerate political opposition of any sort, in practice they treated as legitimate a respectful defense of traditional liberties. This unavoidable acknowledgement of diversity and particularism

4

prevented the monarchy from becoming totalitarian in practice. It also created a space within which lively-minded public figures like Lavoisier could pursue courses of action that combined ideas drawn from both the doctrines of monarchic absolutism and those of republican liberalism.

The 1770s and 1780s, the decades considered in part II, were years of intense contestation in French political life. As the monarch's ministers pressed for further centralization, the defenders of traditional rights and privileges were driven to increasingly extreme forms of reaction. They defended with vigor what they termed their ancient liberties and articulated a distinctive political ideology to support their positions. Spokesmen for the nobility of both the robe and the sword and for the church hierarchy were most assuredly not democrats, but neither were they self-serving aristocrats of the sort the Jacobins condemned when seeking to justify revolutionary seizures of power and property. Most aristocrats opposed the growing power of the centralizing monarchy rather than the idea of monarchy itself, and they were acting on principle as much as in their own behalf when they openly challenged the king and his ministers in a variety of forums. Their views were most strenuously asserted by the *parlements*, the nation's sovereign law courts. Lavoisier, who was himself a member of the Order of Barristers of the Parlement of Paris, was thoroughly familiar with the concerns and claims of this aristocratic opposition. In the end the aristocratic reactionaries made it impossible for the king to govern. By bringing the government to its knees, those opposed to further centralization prepared the way for new forms of governance that struck root during the Revolution.

A third, more radical vision of politics was also being articulated in Paris during the 1770s and 1780s. Its partisans accepted in principle the claim that France is a single nation. They challenged, however, the claim that the king is the sole legitimate representative of all Frenchmen. Rousseau's concept of a people bound together by a "general will" was crucial in this regard, as was his motivating moral dismay over the evils of luxury and the corruption of the aristocracy and the court. Populists and proto-democrats insisted that the nation consists of its people and that they alone have the right to select their representatives. This strain of political thought flourished luxuriantly once the

5

reactionary politics of caste and privilege had brought royal government to a standstill. The ascendancy of this conception of nation and state during the revolution had profound political consequences, some of which are examined in part III.

The monarchic drive toward administrative centralization, and the reactionary and populist responses it occasioned, defined the political context of Lavoisier's life up to the beginning of the Revolution. There was considerable room for maneuver in pre-revolutionary politics and Lavoisier made full use of it when deploying the resources available to him within the Academy of Sciences and in his other administrative activities. When the collapse of royal government broke the mold of political life as he had known it, however, Lavoisier, like all his contemporaries, had to concentrate on staying afloat while being swept along by events. He remained relatively optimistic in the mounting storm and right up to the end continued to be as active in public life as circumstances permitted. His was a life filled with intense and sustained public engagements on many fronts. Its story reveals a great deal about the man, the age in which he lived, and a number of attempts to integrate scientific, cultural, and political developments during an historical epoch of exceptional interest and consequence.

Many scholars have examined Lavoisier's role in the chemical revolution. The following chapters on his science contain much that will be familiar to specialists, yet they also go beyond existing accounts by emphasizing certain themes that infused his scientific achievements with coherence and meaning. The interpretation offered begins with the observation that Lavoisier saw himself as applying the method of experiment, especially as utilized by experimental physicists, to certain unresolved theoretical problems in chemistry. His commitment to the precise use of instruments, to analytic rigor, and to experimental verification of theories constituted a self-conscious, although not original, scientific style. Lavoisier certainly was not the first or only chemist of his time to privilege the techniques of experiment. Rather, like many of his contemporaries, Lavoisier saw the methods of experimental physics as an especially powerful resource that enlightened and public-spirited *philosophes* could deploy with advantage when engaging problems in both science and public administration. At one point, for instance, Lavoisier suggested that enumerating the population of France provided a "thermometer of public prosperity." And he would have applauded, I believe, the author who declared that "experiment, research, calculation are the probe of the sciences. What problems could not be so treated in administration!"[2]

Although thoroughly committed to the instrumental, analytic, and theoretical methods of experimental physics, Lavoisier realized that the science of chemistry could not be transformed simply by making more rigorous use of a well-known methodology. Lavoisier's scientific achievement belongs to chemistry rather than physics because he succeeded in marrying his rhetoric of numbers and his methodology of experiment to significant chemical problems.[3] His mastery of chemistry emerged from a prolonged and strenuous engagement with the investigative traditions and distinctive concepts of chemistry itself. To understand the origins and meaning of that achievement, therefore, one must examine both the style and substance of the research program he pursued.

Early in his career Lavoisier fixed his attention on certain chemical concepts and problems he encountered while studying with the chemist L. C. de La Planche and while attending the lectures of G. F. Rouelle. The research program he constructed to investigate these problems, and the discoveries and theories for which he is remembered, are described below in chapters 3, 4, 6, and 7. Lavoisier's new theories were of central importance to the chemical revolution of the eighteenth century, but the account of his science presented in this biography should not be read as a history of the revolution itself. The chemical revolution was the work of many hands and his allies and opponents reacted to his innovations in ways that are far too complex to chronicle here. Therefore, while this biography will, I hope, help clarify our understanding of the chemical revolution, it does not pretend to provide a comprehensive reinterpretation of that larger event.

Part I

Ambition, Knowledge, and Public Service: 1743 to 1775

1

The Barristers of Paris

Lavoisier celebrated his twenty-first birthday on 26 August 1764. He was spending that summer, as he had spent the preceding years, living quietly and comfortably in Paris with his father, his maternal grandmother, and his aunt. He had just finished several years as a student of both law and science. He completed his legal studies at the beginning of the summer and was looking forward to committing himself to a career. It appears that Lavoisier, who was not given to indecision, already considered himself destined for science. He had made certain scientific observations while a student, but it was not until the summer of 1764 that he undertook a research program specifically designed to earn him a prominent place in the world of French science.[1]

Lavoisier studied law to satisfy his family's expectations. The law was a safe, respectable profession, and had his fascination with science turned out to be nothing more than a youthful enthusiasm, he could have returned to the law at a later date. Lavoisier knew that having access to an honorable calling was a matter of considerable importance to young men in the hierarchical society of eighteenth-century France. His family had done well through its association with the courts in Paris and he had good reason to secure his claim to this part of his

11

inheritance before venturing forth into the less secure world of science.

Lavoisier qualified for the law by completing the required three years as a student in the Paris law faculty; he was awarded a bachelor's degree after the first two years and a *licence* following the third. With these certificates in hand he presented himself as a candidate for admission to the Order of Barristers (Ordre des Avocats) of the Parlement of Paris. He was admitted to this distinguished legal fraternity at an unusually early age in July 1764. Although he had no intention of pursuing a career in the law, he was aware of the social prestige associated with being a barrister; in the marriage contract he signed in December 1771 the first of the corporate affiliations by which he is identified is that of barrister.[2]

Lavoisier had not neglected his scientific interests during his years in the faculty of law. While a law student he studied with several of the prominent scientists who offered both public and private courses in Paris. But a young man hoping to make a name for himself in science had to be concerned about the social organization of science as well as its cognitive content. In the summer of 1764 Lavoisier therefore shrewdly focused his attention on gaining a place in an institution that would provide the kind of support he needed. He was looking for an established corporation or office that would make available instruments and other forms of material support and would also provide a forum to which he could report his findings. It was important to a man of his ambition that this institution be of such high social standing that those who distinguished themselves within it would be recognized by the world at large as leading scientists. The institution he chose was both obvious and perfectly suited to Lavoisier's vision of his future; in the summer of 1764 he began his campaign for admission to the Royal Academy of Sciences in Paris (Académie Royale des Sciences).

The Academy was at that time conducting a public competition on the question of how best to illuminate city streets. Lavoisier quickly decided this topic offered him a perfect opportunity to display his talent as an experimenter and his commitment to using scientific knowledge for the public good. By August 1764 he was hard at work on an extended series of experiments on the construction, fueling, and placement of street lamps.[3] The essay he composed and submitted to the Academy after many months of strenuous effort contains abundant evidence of his dedication, ambition, and talent, and the Academy of

Sciences eventually recognized its merits by awarding him a special medal. This essay also played an important part in the sustained campaign that four years later culminated in Lavoisier's being elected to the Academy of Sciences. In science as in the law, Lavoisier gained admission to a prestigious corporation at an early age and long before distinguishing himself as a practitioner. In addition to being immensely talented and energetic, he enjoyed the considerable advantage of starting out at the top.

Lavoisier's father was not a Parisian by birth. His family came from Villier-Cotterets, a town about fifty miles northeast of Paris, in which Antoine later spent many a happy boyhood summer. In the century preceding Antoine's birth the Lavoisiers had advanced their fortunes through minor positions in the king's service and in trade. Antoine's father, Jean Antoine, made the move to Paris when sent there by his mother's brother to study law. This uncle, Jacques Waroquier, was a solicitor (*procureur*) at the Parlement of Paris. Waroquier had no immediate descendants and when he died in 1741, his office, which he owned outright, and his house in Paris were passed on to his nephew Jean Antoine.

A year later, in 1742, Jean Antoine married Émilie Punctis, one of two daughters of a well-established bourgeois family in Paris. Émilie mother's family had been successful Parisian butchers and the greater part of the Lavoisier family fortune, which was considerable, passed down the female line. Antoine later used the money he received from his mother's estate to purchase a share in the Tax Farm. Émilie's father was, among other things, a barrister at the Parlement of Paris and he no doubt helped Jean Antoine Lavoisier move from the less distinguished post of solicitor to membership in the Order of Barristers. Antoine, who remained close to his father until his death in 1775, clearly appreciated the ways in which the law, and more especially membership in the Order of Barristers, had helped the Lavoisiers improve their position in French society. It should also be noted that Antoine's future father-in-law, Jacques Paulze, with whom he worked as a tax farmer, was also a member of the Order of Barristers.[4]

Émilie Punctis brought to her marriage a dowry of 17,000 livres and two minor annuities; Jean Antoine's personal assets at the time of their wedding were valued at 42,000 livres. These were substantial sums at a time when the church expected a curé to get by on 700 livres per year and the state considered an annual salary of 2,400 livres for a professor at the Military School in Paris a handsome stipend.[5] Thus while the Lavoisiers were not wealthy by the standards of France's great aristocrats and merchants, they had much to be thankful for and, should misfortune strike, much to lose. Antoine, the first child, was born in the second year of this marriage; his sister, Marie Marguerite Émilie, who was to die when only fifteen years old, was born two years later. This completed the family, for the children's mother died when Antoine was only five. His father, although only thirty-three at the time of her death, never remarried. In their bereavement the family turned inward for strength.

Antoine's maternal grandmother was widowed the year before the death of her daughter, and his father and grandmother responded to their double loss by establishing a common household in the grandmother's home. Another of the grandmother's daughters, Constance Punctis, was twenty-two at the time and joined them to look after the two children. Antoine spent six years in this doting ménage before beginning his formal education as a day student at the Collège des Quatre-Nations, popularly known after its founder as the Collège Mazarin. In that same year, 1754, his grandmother Punctis received her one-third share of the estate of 137,000 livres left by her recently deceased father.

The death of Antoine's sister six years later further increased the family's inwardness. Grandmother Punctis's fondness for quiet and solitude grew stronger as she aged, and only a few intimate friends were admitted to the house on rue du Four-Saint-Eustache. Young Antoine, with all of Paris to divert him, evidently appreciated the island of domestic calm to which he returned daily. When he married in 1771, three years after his grandmother's death, Lavoisier and his bride settled down in a nearby house that his father purchased for them.

In what ways was Antoine's personality shaped by his family's circumstances and attitudes? We can hazard a few guesses on how the Lavoisiers' intimate family life may have reinforced certain of the young man's inherent characteristics, but the available evidence is too insubstantial to allow us to say anything definitive. We can speak with somewhat greater confidence, however, about the social position of the Lavoisiers and the ways in which Antoine's attitudes and ambitions reflected his family's social circumstances. The Lavoisiers, while well-connected and financially secure, were well aware they were relative newcomers in Paris. Proud of their success, and especially of their service to the king and the law courts, they expected the family to prosper and advance. Yet they also knew that the society in which they lived was intensely protective of status and privilege and that the presumption of anyone seen to be overreaching his station would be punished unmercifully. The family encouraged ambition, but it also recognized the need for caution, respect, and restraint. There were ample opportunities for further improvement, but fate and those who controlled the higher reaches of society could at any time turn malign. To succeed in the contested politics and culture of the time, one had to be shrewd and reserved as well as industrious and intelligent. This view of the world differed considerably from that of the bounding revolutionaries with whom Lavoisier had to contend toward the end of his life. When compared to them Lavoisier appears conservative, even though he was no less passionately committed to improving public life in France.

Lavoisier's sense of his responsibilities as the sole surviving son, and his determination to meet the high expectations of the proud and pious elders among whom he was reared, no doubt encouraged conduct that occasionally bordered on the obsessive. When he was nineteen years old, for instance, Antoine decided to investigate the effects of diet on health by ingesting nothing but milk. As a concerned friend commented, such behavior exhibits the kind of unreasonableness so characteristic of intellectuals: "your health, my dear mathematician, is like that of all men of letters in whom the spirit is stronger than the body. So be sparing in your studies and accept that another year on earth is better than a hundred in the memory of man."[6] Lavoisier was again prepared to put his health at risk for the cause of science when performing experiments on street lighting. He proposed to shut himself up in a completely dark room for six weeks so that he could better judge slight differences in the intensity of illumination, but we do not know whether he actually did so.[7]

A third example of such behavior occurred in 1768, when Lavoisier was trying to determine whether water could be transformed into earth by repeated distillation. He heated a sample of water in a pelican, a device that returns the condensed distillate to the liquid being distilled, and kept it at a temperature just below boiling point for 101 consecutive days, an effort which required constant vigilance.[8] Additional indications of Lavoisier's youthful intensity can also be found in the manuscript notes preserved in the archives of the Academy of Sciences. He had an insatiable appetite for recording meterological information, especially barometer readings, and for analyzing samples of water taken from wells. Clearly he was a driven young man. Having decided to distinguish himself in the world of science, he threw himself into the task with unrestrained energy and concentration. Throughout his life he exhibited an enormous capacity for detailed and often tedious work, a facet of his personality that might well have been rooted in a consuming inner need to succeed.

But what counted as success for Lavoisier? The question is not as simple as it seems and before attempting to answer it we need to make some preliminary distinctions. The first of these differentiates between science defined as a body of knowledge and the scientist considered as an historical actor. Science itself is normally thought of as those facts and theories that provide the best available account of nature. The task of doing science is the business of specialists; they are responsible for determining what constitutes good science and for assessing the empirical and theoretical claims of their colleagues. When science is conceived of in this way, arriving at a consensus on what constitutes success in science seems fairly straightforward – success in science is determined by the specialists who decide which are the best theories, which experimental discoveries support these theories, and which scientists deserve credit for advancing our understanding of nature. Those who are judged to have excelled according to these criteria are today awarded Nobel prizes.

Lavoisier's achievements, when evaluated by these standards, fully sustain his reputation as one of the most successful figures in the history of science. But in addition to treating science as an end in itself Lavoisier, like all scientists to a greater or lesser extent, saw science as a means to other ends as well. As an individual making his way in the world, he brought to the pursuit of science a vision of success that was both broad and complex. Experimental and theoretical success were not the only goals he had in mind; indeed, he often treated his cognitive successes as steppingstones on the way to more comprehensive social and cultural goals. We therefore need to distinguish between two different measures of success in Lavoisier's science. One asks what notable experiments and theoretical innovations he contributed to the science of his time. The other asks to what extent he reached the social and political goals that were an integral part of his conception of science. This latter question is one which the biographer in particular must engage.

Another distinction we need to keep in mind is the difference between the organization of society in late eighteenth-century France and the social organization of the most developed nations at the end of the twentieth century. Dramatic differences in this regard, and the ones that are most easily overlooked, are political and ideational rather than economic and technological. Two hundred years of preoccupation with the problems of democracy, equality, and individual rights stand between our vision of what constitutes the proper order of society and the ideas of Lavoisier and his contemporaries. Their society, which revolutionary zealots castigated as "the old regime," had evolved a complex hierarchical structure, and its many privileges and obligations were allotted according to one's membership in various social orders rather than to individuals as a matter of right.

A petition the Parlement of Paris addressed to the king in 1776 provides a concise description of the corporate and hierarchical assumptions of fundamental importance in eighteenth-century society.

> All your subjects, Sire, are subdivided into as many different corporate bodies as there are different callings [*états*] in the Kingdom: the clergy, the nobility, the sovereign courts, the lower courts, the universities, the academies, the trading companies, all of which produce, in every part of the State, extant corporations that one can consider as links in a great chain, the first of which is in the hands of Your Majesty, as head and sovereign administrator of all that comprises the body of the Nation.[9]

This corporatist and decentralized social structure was vigorously defended in principle as well as in the name of tradition. The main question at issue was where sovereign authority should reside; comparisons with Great Britain were considered especially instructive. The British Crown, having experienced revolution and regicide in the seventeenth century, feared rebellion; French jurists, after nearly a century of increasing administrative centralization, feared despotism. The central difference is nicely captured in the following exchange. In 1749 Charles Yorke, son of the British Lord Chancellor Hardwicke, told the French political philosopher Montesquieu that his father had remodelled Scottish law following the Jacobite Rebellion of 1745 so as to deny the chiefs of the Scottish clans their traditional power to dispense justice. Montesquieu, who had been a magistrate in the Parlement of Bordeaux, questioned the wisdom of this reform, for he believed the traditional rights of the Scottish chiefs provided "a barrier against the Crown to prevent the monarchy from running into despotism."[10] Yorke, seeing only faction, feared rebellion; Montesquieu, seeing only absolutism, feared tyranny.

Lavoisier fashioned his career in a society of orders, each of which had distinctive duties and privileges. He had fully internalized his family's views on what constitutes social success and he was guided by those beliefs at every stage of his life. The world of the barristers of Paris provided his point of departure; the world of science, broadly conceived, constituted his primary arena for public action. To succeed as a scientist he had to produce significant experimental discoveries and theoretical innovations. But his larger goal was to ensure that that kind of success would also contribute to his advancement in the hierarchical and intensely competitive society of eighteenth-century Paris. He could, after all, have satisfied his curiosity about nature by cultivating an avocational interest in science while pursuing a legal career. That he did not do so indicates that, in embarking on a scientific career, Lavoisier was openly committing himself both to the pursuit of scientific knowledge and to an ambitious program for social advancement through science.

Barristers presented cases before the Parlement of Paris, France's preeminent sovereign court.[11] The magistrates before whom they pleaded vigorously defended their privileges as members of the nobility of the robe; the barristers occupied a less clearly defined position in the upper

reaches of society. They certainly basked in the aura of privilege surrounding the magistrates, yet being a member of the Order did not in
itself confer nobility. Theirs was nonetheless a gentlemanly calling.
Barristers were often said to possess a kind of personal nobility and
many of them lived like aristocrats. Membership of the Order of Barristers did confer certain privileges within the law; members were exempt
from all personal service, including service in the militia. The barristers
were especially vigilant in excluding from their ranks those who held
other offices or engaged in public activities that did not accord with the
high status of their calling.

The Order was a self-governing association, not a corporation established by letters patent from the king. The barristers set their own rules
and controlled admission to their Order, and the office of barrister,
unlike that of solicitor, was not venal. Institutional autonomy of this
sort was unusual in eighteenth-century France. The barristers took
their responsibilities seriously and forthrightly defended their independence against the centralizing tendencies of the monarchy. This
strategy was not in fact very risky, for the Bourbon monarchs largely
left the legal profession to its own devices, their few attempts to regu-
late legal education being ineffective. The Order could therefore assert
with some plausibility that within the world of the law it was the
beleaguered champion of high standards.

In the eighteenth century the instruction offered by French law
faculties was notoriously bad. Laxity and corruption were rampant and
students could get the certificates they needed by paying fees rather
than attending lectures. Lavoisier was thus simply conforming to current practices when he devoted much of his time during his years as a
law student to the pursuit of science. Three years of legal studies were
normally required before one could apply to be a barrister. Given the
slackness of the law faculty, the Order had assumed responsibility for
ensuring that those it admitted as members possessed suitable personal
qualities and sufficient legal knowledge. Candidates were therefore
required to pass through a four-year probationary period before being
admitted to full membership. Evidently Lavoisier was either excused
from this requirement or had already begun his probationary period
when he enrolled as a law student. Although special arrangements
were no doubt made for well-connected young men of spotless charac-

ter and demonstrable talent, the Order of Barristers could maintain with some justice that it was a bastion of high standards and responsible self-governance in a profession that had otherwise drifted into disarray.

The rigor with which the barristers controlled entry to their ranks was not matched by a comparable vigor in the practice of law. Many members joined the Order solely to establish themselves as gentleman; it was said that roughly half of the barristers did not engage in legal practice. Most of the barristers lived in the Marais section of Paris, where the Lavoisiers also lived before Antoine's mother's death. Those who presented cases in the nearby buildings of the Parlement were not noted for their ambition. Barristers usually handled only one case at a time and did not devote a great deal of attention to their legal duties. Such leisurely work habits were quite acceptable in a calling that approached the status of nobility. However, careers of this sort offered few attractions to those who, like Lavoisier, were seeking challenges that would reward intense exertion. Lavoisier also doubtless knew that the barrister's profession was not especially lucrative. There were more than enough barristers for the work to be done and the most prominent of them seldom attempted to enlarge their practices. One did not become a barrister to make money; the important rewards were social, not economic.

Barristers, like most office-holders, were more concerned with their status than their incomes, and in the second half of the eighteenth century they had many opportunities to defend their rights and privileges. In the seventeenth century Louis XIV had seized control of French political life. He insisted that by the grace of God and through administrative action the monarch alone represented the nation as a whole. The established estates and orders accepted this assumption of authority because he promised to protect their privileges. He also invited those members of the nobility of the sword who wished to participate in public life to join in the elaborate rituals of his court. His successor, Louis XV, inherited the crown as a minor and respect for the monarchy began to wane as strong ministers grasped the reins of power.

One of the traditional duties of the French *parlements* was to register and interpret royal edicts, and by the 1750s, as they grew ever more dissatisfied with the reign of Louis XV, the parlementary magistrates were becoming increasingly militant in defending their constitutional rights. Disagreements over the distribution of political power within the nation became acrimonious. The demand for state financial, military, and administrative services was expanding while the prestige of the monarchy was declining. Tensions were escalating into open confrontations. While the barristers no doubt would have preferred to cultivate their gentlemanly leisure, they found themselves caught up in the political whirlwinds that in the end brought on a revolution.[12]

The *parlements* defended their rights by appealing to past practice and to the writings of Montesquieu. In 1753 the Parlement of Paris refused to register royal edicts supporting the Pope's authority over the Church in France. The Grand Remonstrance it forwarded to the king summarized the magistrates' position forthrightly.

> If there are times when the court's unshakeable attachment to the laws and to the public good seems to accord ill with a limitless obedience, then it would be wrong . . . to forget . . . what the Parlement told the sovereign in 1604: if we disobey by serving you well, then the Parlement is frequently guilty of this fault. When there is a conflict between the king's absolute power and the good of his service, the court respects the latter rather than the former, not to disobey but in order to discharge its obligations.[13]

The king refused to accept this reading of the situation and rejected the remonstrance. The Parlement responded by going on strike; the king retaliated by exiling the magistrates from Paris and setting up a royal chamber to carry out their functions. This body lacked traditional authority, however, and the barristers boycotted it. A compromise was then arranged, but when the magistrates returned in 1754, they treated their survival as a great victory and felt strengthened in their defense of their traditional rights.

The scenario enacted in 1753–4 was replayed repeatedly during the final two decades of Louis XV's reign, but always with heightened drama and increasingly dire consequences. When the Seven Years War ended in 1763 it was thought that certain war taxes would be removed. When Bertin, the Controller-General of Finance, drafted an edict continuing the taxes, the Parlement of Paris reacted with fury. The king soon buckled, the edict was withdrawn, Bertin was dismissed, and a member of the Parlement was appointed as his successor. In 1764, the year Lavoisier joined the Order of Barristers, the Parlement of Paris used similar tactics to force the king to suppress the Jesuit order throughout France.

This mounting drama of conflict and confrontation reached its climax in 1771. In 1766 the king, having decided to take a stand, declared in a *lit de justice*, an extraordinary judicial session in which the king asserted his legal authority, that he would no longer tolerate parlementary obstruction. He rejected the *parlements'* assertions of authority and insisted on the Bourbon claim to absolute power over all aspects of the law. But the *parlements* had enjoyed too much success to be easily overawed and they fortified their positions with elaborate defenses of their rights. When the king rejected these appeals to precedent and tradition, the issue was joined. As Voltaire commented, "this astonishing anarchy cannot last, either the Crown must re-assert its authority or the Parlements will gain the upper hand."[14] In January Chancellor Maupeou, the king's minister for legal affairs, seized the offices of the Parlement of Paris and sent the magistrates into exile. The king's intention was to resolve the situation once and for all by completely eliminating the Parlement of Paris.

Maupeou's suppression of the Parlement of Paris created an awkward situation for the barristers. He carefully avoided disturbing their rights and privileges before asking them to serve the new court created to replace the Parlement. When the barristers demonstrated their solidarity with the exiled magistrates by refusing to have anything to do with the new court, Maupeou was forced to try another approach. He created a corps of a hundred Avocats *du* Parlement to perform the duties previously reserved for members of the Ordre *des* Avocats. Since there was no way of knowing whether the old court would ever be reestablished, the existence of this new corporation constituted a serious threat to the social position of the barristers. The old Order of Barristers did not surrender, however, and the struggle continued. The confrontation was only resolved by the death of the king in 1774 and the dismissal of Maupeou. The new king, Louis XVI, soon recalled the former magistrates, thereby ensuring the survival of the Parlement and the continued vitality of its claims to authority.

Those associated with the *parlements* were inclined to defend the traditional constitution, the received social order, and regional and corporate privileges. They were, conversely, unmoved by pronouncements favoring political centralization and appeals to universal values, such as those offered by many enlightened theorists. The barristers, while proud of their achievements in the world of letters and especially in the writing of histories, exhibited little interest in Diderot's and d'Alembert's *Encyclopédie*.[15] As Mercier wrote in his *Tableau de Paris*, the barristers usually preferred "the authority of old books to the authority of reason."[16]

Unlike the theorizing *philosophes*, whose prescriptions for reform were routinely coupled to declarations about the order of nature, the parlementarians took the legal tradition and the historical record as their points of reference. In the polite world of eighteenth-century culture they preferred an ironic skepticism to rationalism, and the rhetorical style of literary culture to speculations founded on observations of nature. The parlementarians were seriously concerned with justice, political order, and public administration, but they were unmoved by arguments founded on claims to universal rights and deductions from natural laws. Like other men of leisure, they were frequently amused and informed by scientific studies, but, as social theorists and political reformers, they began with human history rather than natural history.

The cultural world of the barristers was thus marked by certain patterns of thought that helped shape the young Lavoisier's mental landscape. The impress of these beliefs can be traced throughout his career and scientific achievement. Although he excelled as a theorist, Lavoisier was always a positivist, in the eighteenth-century sense of that term, in that he showed little interest in theories that were not well-grounded in a dense bed of facts. While fully appreciating the need for inductive generalizations in science, he showed almost no interest in comprehensive systems of natural philosophy. He adhered rigorously to a systematic methodology when reasoning and avoided the kind of speculative system-building favored by the older *philosophes*. He seldom went beyond proposing limited generalizations that could be verified by experiments. Lavoisier concentrated on the study of nature rather than the study of the human past, but unlike many of his fellow reformers, he was careful not to slide from descriptive statements about relations in nature to prescriptive statements about social and political relations.

Lavoisier chose not to spend his life in the social and cultural milieu of the barristers, but he never rebelled against it or renounced this aspect of his intellectual heritage. Although he broke with the barristers' commitment to France's feudal constitution and tradition of decentralized governance, he did so as a pragmatic reformer, not as someone seeking to impose an abstract system on a recalcitrant nation. He found his pleasure and made his mark by studying nature rather than human history, yet he never deified nature or treated politics as a derivative enterprise. His devotion to his nation was deeply rooted in his familiarity with its history, and his unshakeable commitment to stoic resolve and public *politesse* reveals how thoroughly he had internalized the values instilled in him during his youth. Culturally and socially Lavoisier was and remained a barrister; by nature and by choice he was and became a man of science.

2

The Republic of Science

On 15 May 1753 Jean Antoine Nollet delivered his inaugural lecture as Professor of Experimental Physics in the University of Paris. The king had created this position especially for Nollet and the inaugural lecture was delivered in the large amphitheater of the College of Navarre, an auditorium for 600 that had recently been constructed for the lecture course the new professor was to offer. The ceremonies surrounding Nollet's installation also marked the culmination of over two decades of growing interest in experimental physics in Paris, an interest that Nollet himself had energetically and skillfully encouraged.[1]

Nollet, whose parents were provincial peasants, carefully cultivated his reputation as a prominent scientist. He was best known as an immensely popular public lecturer. In the late 1740s he, like several other experimenters, used the recently discovered Leyden Jar to demonstrate the power of electricity. In the king's presence Nollet shocked a line of 180 soldiers, who responded with gratifying alarm, and he later repeated the demonstration with 200 Carthusians in their Parisian monastery. It is no wonder that his lectures on experimental physics were much discussed in polite society, or that the course he gave in 1760 attracted an audience of 500.[2]

Nollet was indeed an effective showman, but he was also an accomplished theorist and a leading scientific spokesman. He was an active and respected member of the Academy of Sciences, and in France between 1745 and 1752 his theory of electrical action was universally considered the best available.[3] Among his colleagues Nollet was also known for his unwavering politeness and unfailing consideration for others. In 1743 at Versailles he tutored the queen and the dauphin, the king's son and heir apparent, and he was a welcome figure in both the salons of the *philosophes* and the monarch's court. The king was thus rewarding Nollet's scientific achievement, his service, and his popularity when he appointed him to the first French professorship in experimental physics.

Nollet titled his inaugural lecture "The attitudes and attributes required to achieve progress in the study of experimental physics." In the course of describing these attitudes and attributes he characterized the community of science as a republic.

> One should be taught early on that those who cultivate the sciences (*les Sciences*), no matter what part of the world they live in, are members of a single republic (*République*). They owe each other the mutual respect one expects of fellow citizens (*Concitoyens*). Members of the republic of science work together for enlightenment and engage in only those honorable forms of competition (*émulation*) that spring from the desire to surpass one another; they do not attempt to ridicule or confuse their compatriots.[4]

Lavoisier attended Nollet's lectures in the early 1760s and his taste for experimental physics and his appreciation of science as a form of polite culture were strengthened by Nollet's inspired advocacy. When viewed in retrospect, Lavoisier's investigations of air, a substance that hovered on the borderland between physics and chemistry, appear quite similar to Nollet's investigations of electricity, a subject that from a disciplinary point of view floated as freely as air. Lavoisier also found Nollet's republican image of science congenial and he especially welcomed the way it linked a citizen's commitment to the public good and a politely competitive investigation of nature. Science, as Nollet described it, appeared to be one of those rare enterprises in which an ambitious young man could do well individually while doing good socially.

26

In retrospect again we can see that Nollet, in his scientific career, had achieved much that Lavoisier as a youth dreamed of attaining. While the two men came from very different backgrounds and pursued their careers in different generations and circumstances, there are striking similarities in their public lives as scientists. Nollet, speaking from on high at the peak of his career, described in elevated terms the personal characteristics required for success in the science, experimental physics, that had rewarded his efforts so richly. Lavoisier must have found the themes developed in this academic sermon compelling when he heard them repeated in Nollet's public lectures on physics. The images Nollet employed and the values he commended were precisely those Lavoisier accepted as his own while preparing himself for entry into the republic of science.

The republic of science, like the republic of letters, was a special kind of community. Today we would probably call it a network, for it was above all else a community of discourse. Members of the republic of science investigated nature and reported their findings to each other; they arranged those findings in a systematic manner and interpreted their meanings; they evaluated interpretations proposed by others and defended their own conclusions. They did all this in open discussion with other members of the republic. Discourse and dialogue were the central communal activities of the republic; the tacit and consensual agreements that made such verbal interaction possible were its fundamental laws.

Those who wished to participate in the activities of the republic were expected to abide by its standards of conduct. The republic's distinctive code of behavior, rather than the specialized languages of the different sciences, gave form and direction to the community of science. The republic of science, like the ancient Roman republic and the post-Renaissance republic of letters, rested ultimately on a set of shared beliefs about the duties of its citizens and the ways in which they should relate to one another.[5] The image of science as a republic rendered intelligible its social organization, its patterns of authority, and the public role of the scientist, and it did so in a way that practicing scientists in the latter decades of the eighteenth century found persuasive and inspiring.

Many different institutions provided support for the republic of science, although none of them had been established primarily to encourage science for its own sake.[6] In pre-industrial France the sciences, like orchids in a tropical forest, flourished by drawing sustenance from institutional trees firmly rooted in the nourishing soil of national service. The various academies, universities, museums, and state-supported industries that employed scientists existed not to support science as such, but to meet more immediate needs considered essential to the culture and economy of the nation. This was true even of the Royal Academy of Sciences in Paris, the foremost scientific institution in eighteenth-century France.[7]

While the Academy of Sciences' specific charge was to promote the advancement of scientific knowledge, its more general purpose, as one of the several academies established during the reign of Louis XIV, was to subordinate the cultural activity called science to the monarch's patronage. To capture and domesticate the various branches of what we now call high culture, the king agreed to endow the academies with elevated social status and appropriate means of support. In accepting his patronage, scientists and other academicians agreed to use their expertise in ways that would glorify the monarchy and help it fulfill its obligations to the nation. The Royal Academy of Sciences included among its members leading scientists and provided them with stipends and other forms of support. Yet the king certainly had not intended to sanction republican values or encourage republican practices when he legitimized science by granting its Academy royal status.

A look at the various audiences for science in eighteenth-century France reveals the complexity of the social arrangements that sustained scientific activity. The audience with the greatest political interest in science consisted of the king and his ministers, their concern being to ensure that science remained firmly within the established culture that sustained the absolutist claims of the Bourbons. There was also a growing audience for science in polite society. Lavoisier frequently addressed this audience, if somewhat indirectly, by summarizing his research at public meetings of the Academy and by writing books and articles that were accessible to those having a serious if amateur interest in chemistry. But he always found himself too busy with other commitments to offer a comprehensive public lecture course or to complete an introductory chemistry textbook, and at no stage in his life did the progress of his career depend fundamentally on the approbation of those having only an amateur interest in the study of nature.

The dominant audience within science itself consisted of the core "public" of acknowledged masters that formed a powerful if somewhat veiled high tribunal. The members of this tribunal were those scientists whose judgments carried great weight and whose assessments largely determined what was to be considered good science. Their authority, while often formalized and strengthened by institutional appointments, was ultimately founded on consensual recognition of their empirical and theoretical achievements as individuals. Responsibility for determining what constituted the cognitive content of science thus remained in the hands of scientists, even though the institutional organization and cultural legitimation of science were inextricably linked to the environing worlds of politics and society.

Lavoisier, like Nollet, was honoring core values of his age when he used the language of republicanism to describe his aims as a scientist. Science needed leaders who, like exemplary Romans, were morally upright and devoted to the public good. Lavoisier and his contemporaries did not think of science as esoteric knowledge, nor did they consider a life in science a priestly vocation. Science for them was an integral part of the secular culture of the Enlightenment. The Burgundian Lazare Carnot, trained as an army engineer and an elected member of the Legislative Assembly, later conformed to this image of the citizen's duties when, like a Roman farmer laying down his plow to take up shield and spear, he put aside the scientific studies and political activities that had occupied his time and turned to organizing the French armies during the revolutionary wars.[8]

This image of science also helped Lavoisier's younger collaborators, Pierre Simon Laplace and Claude Louis Berthollet, fashion careers of enormous prestige and power in post-revolutionary France. Having survived the Terror, they received important cultural appointments under Napoleon, that supreme exploiter of Roman symbols. They, like Lavoisier, saw science as a public and national enterprise, not a private amusement or a reclusive search for the inner workings of God's creation. The republican image of science thus provided Lavoisier and his age with a vision of extraordinary breadth and power.[9]

But before Lavoisier could become a notable citizen-scientist he had to convince his elders that he possessed the personal virtues and intellectual abilities expected of a republican. To gain the approbation of this most demanding public he had to demonstrate in appropriate forums that he had mastered the elements of science and was capable of making notable contributions in areas of research that were of interest to the republic. Knowing this, he set out with characteristic energy and intensity to acquire the knowledge and techniques of importance in those fields of science to which he was especially attracted.

Lavoisier spent his first six years at the Collège Mazarin studying the traditional curriculum in humanities. In the final year, devoted to the rhetorical analysis of classical texts, his achievements were recognized by the award of prize books in French, Latin, and Greek. In the same year, and in the month in which he turned seventeen, Lavoisier also won the second prize in a literary competition open to all students enrolled in the secondary schools of the University of Paris.[10] After completing the humanities curriculum in the summer of 1760 he began the three-year course of study in mathematics and philosophy taken by students preparing for the baccalaureate degree. The first of these years was devoted to mathematics, the instructor being the distinguished astronomer Nicolas Louis de Lacaille. Apparently it was while studying with Lacaille that Lavoisier first discovered the delights of science.

Lacaille, a modest and rather severe descendant of a distinguished provincial family, is best known for having spent three years in the southern hemisphere observing and cataloging nearly 10,000 stars.[11] An observer and a calculator rather than a speculator and a theorist, Lacaille impressed on Lavoisier the fundamental importance of having good instruments and using them effectively. He first made a name for himself as a member of the team that settled the Cartesian–Newtonian dispute over the shape of the Earth; they did so by surveying an accurate geodesic baseline in France. In 1738, when only twenty-six years old, he was appointed to the mathematics chair at the Collège Mazarin. A serious instructor, he produced a series of widely used textbooks. The first, on mathematics, was published in 1741, the same year in which Lacaille became a member of the Academy of Sciences. During the following decade and a half he pursued a rapidly expanding program of observation that culminated in his mapping of the southern sky.

In 1745 Lacaille returned from his years in the southern hemisphere. Although celebrated for his achievement, he chose to return to the routines of the observatory and the classroom. He continued to reduce his observations and revise his texts while teaching the subjects Lavoisier began studying with him in the autumn of 1760. Lavoisier was evidently impressed by Lacaille's instruction, for after completing the mathematics class he began a second year of study with him as a private pupil in his observatory. Lacaille's health had been broken by years of nocturnal observing, however, and he died in the winter of 1762, when only forty-nine years old.

While studying with Lacaille Lavoisier was also introduced to the science of chemistry by L. C. de La Planche.[12] La Planche may have offered chemical instruction at the Collège Mazarin, but it seems more likely that Lavoisier attended the courses he offered at the Society of Pharmacists (Compagnie des Apothicaires de Paris) and at his private laboratory. After completing the first year of the philosophy curriculum, however, Lavoisier decided to qualify for the law, a decision that obliged him to leave the Collège and enroll in the faculty of law. He continued to pursue his scientific interests, however, by attending public lectures, enrolling in private courses, and engaging in supervised field work. La Planche had clearly aroused Lavoisier's interest in chemistry, for during the three years he spent studying law, Lavoisier attended three courses taught by Rouelle. These included the flamboyant and popular lectures offered to the public at the King's Garden (Jardin du Roi) as well as the private course Rouelle taught to more advanced students in his personal laboratory.

Lavoisier, while a law student, also attended Nollet's public lectures on experimental physics, where he found the lessons taught by Lacaille reinforced and extended. He also acquired a firm grasp of natural history under the tutelage of two of France's most eminent specialists. Bernard de Jussieu taught him about the world of plants as they tramped together on collecting trips in and around Paris; the geologist Jean Étienne Guettard taught him how to study the Earth. Lavoisier was soon captivated by the analysis of minerals and waters, perhaps

31

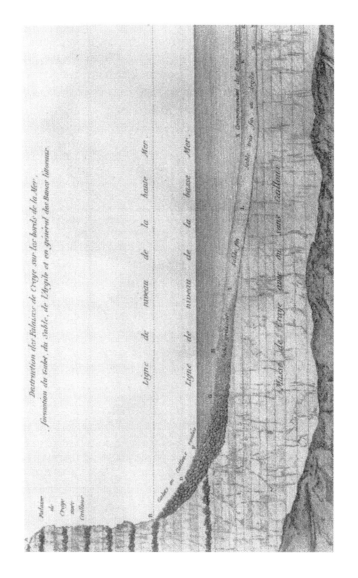

1   An engraving illustrating Lavoisier's theory of coastal erosion and the formation of geological strata.

because he could apply to these subjects the concepts and techniques he was mastering in experimental physics and chemistry. Guettard doubtless encouraged him in this approach, for he too had long been interested in the chemical analysis of minerals and rocks.[13] By 1763 Lavoisier had begun his own mineral collection and had drafted a brief account of the natural history of his family's home town of Villier-Cotterets.[14]

Guettard advised Lavoisier as he puzzled over the formation and nature of the Earth: the Lavoisier family enjoyed Guettard's company, perhaps because as barristers they approved of the gruff bachelor's fervent Jansenism.[15] For over fifteen years Guettard had been traveling widely in France while collecting information for a projected mineralogical atlas of the kingdom. Lavoisier became his enthusiastic assistant in this project, which happily combined the pursuit of natural knowledge with a concern for the public good, and he was soon actively exploring a variety of geological problems. Master and pupil became collaborators and their project was shortly thereafter given official sanction and support. In 1766 Henri Bertin, Minister and Secretary of State for Mining, commissioned them to prepare a geological survey of France. Work on the survey continued for many years and helped focus Lavoisier's investigations as he moved through his scientific apprenticeship.[16]

Once he had gained a preliminary command of the subjects that interested him, Lavoisier concentrated on gaining admission to the Academy of Sciences. The financial support provided by the Academy was not its primary attraction; Lavoisier was, after all, a wealthy and well-connected Parisian who could have funded his own research, had that been his foremost concern. What Lavoisier sought, and what the Academy offered, was an institutional affiliation that provided highly visible political legitimation for the pursuit of science. He did not wish to spend years in remote locations or obscure laboratories while building up a record of achievement that would be rewarded by an offer of membership in the Academy. Lavoisier's campaign for admission to the Academy began before he had distinguished himself in any way other than as a student. He was not, however, merely seeking the social distinction associated with being an academician. He also wanted explicit political recognition that science is an enterprise of great national importance.

The Academy of Sciences, like the other royal academies, served both its subject, in this case science, and the state.[17] As an institution of the state it existed to glorify the monarchy by tightly binding together the king and the scientific elite. This conjunction helped confirm that the king alone represented the nation as a whole. The Academy served the scientific elite by cloaking their enterprise in a sanction of such magisterial authority that they were free to act in public as if they were completely above politics. When taken together, the royal academies created an official culture that successfully dominated public discourse on topics such as language and literature, history, science, painting, music, theater, and architecture. When the monarch turned his gaze on such royal academies as the Académie Française, the Académie des Sciences and the Académie des Inscriptions et Belles-lettres, he had every reason to be satisfied with the return on investment.

Each of the royal academies provided its members with forms of support appropriate to its branch of culture. Stipends for members, rooms in which to meet and work, public sessions and exhibits, the right to authorize publication – these were important if fairly obvious benefits of royal patronage. Less visible, but of fundamental political importance, was the way in which each academy created a protected sphere within which the members controlled their own affairs when displaying the results of their efforts and assessing the work of colleagues. Thus the Academy of Sciences in practice conducted its affairs according to two modes of operation, each of which conformed to a distinctive style of political discourse.

In its internal activities the Academy acted as a republic, with influence gravitating to those who made the most significant contributions to the advancement of science. This arrangement ensured that responsibility for determining scientific merit would remain in the hands of the most informed and discriminating public for science, the members of the Academy itself. But in its relations with the world outside the Academy acted with monarchal authority. Its dependence on royal patronage was acknowledged, its obligation to report on the problems and proposals referred to it was accepted, and its authority in determining questions of scientific truth and technical utility was insisted upon and defended. The politics of academic life were therefore fairly complicated. But Lavoisier, who always showed greater interest in getting things done than in striking public poses, was not put off by the complex and contentious politics of institutional life in eighteenth-century France. He knew precisely where the Academy of Sciences stood in the social, cultural, and intellectual order of his day and he set about making his career within the institutional world as he found it.

The Academy of Sciences had been formally established in 1666, when Louis XIV's chief minister Colbert created an official institution designed to bring under the king's protection what had until then been a self-constituted and slowly developing community of scientists. Lavoisier, in an *éloge* of Colbert drafted in 1771, used extravagant yet revealing imagery when praising him for establishing the royal academies.

> He called together learned men and artists of all types and set them up in little republics whose vitality endures from age to age. These institutions, the finest gift mankind has ever received, are monuments erected against ignorance and barbarism. These bodies, being endowed with an active power, not only preserve from age to age the initial impulse given them by a great minister, their active power also overwhelms any opposition arising from ignorance, superstition, and barbarism. I happily compare these institutions to the great bodies that revolve above us and to which the creator of all things gave an initial motion that they have preserved since the beginning of the world.[18]

The organization and internal structure of the Academy reflected the preoccupation with status and patronage that permeated French society. The highest class of members consisted of twelve honorary academicians, but the real work was left to approximately fifty members appointed to the six scientific sections. Three of these sections addressed aspects of what were termed the mathematical sciences – geometry, astronomy, and mechanics – and three were concerned with the so-called physical sciences – anatomy, chemistry, and botany. Each section had three pensioners, the most senior rank for working members and the only appointments that were salaried, two associates, and two adjuncts, the most junior members. Supernumeraries were occasionally appointed as well. Academic salaries averaged about 2,000 livres a year, a decent salary by the standards of the time but not enough to maintain a comfortable bourgeois style of life in Paris. Many pensioners therefore held other teaching or governmental positions or drew on other sources of income.

In its internal governance the Academy operated much like a modern university department.[19] Rank was acknowledged and accompanied by certain privileges, yet there was also a strong sense of collegiality and collective responsibility. Evidence and argument were the only permissible instruments of public persuasion, achievement was assessed and rewarded, and claims not supported by scientific reasoning were ignored when not scorned. It was understood by the king's ministers and the academicians that the republic of science could flourish only if the liberties of its citizens were respected and defended. Both parties therefore worked to maintain a balance between monarchal and republican expectations within the institution.[20] As Lavoisier pointed out in 1793, in an unsuccessful last-ditch defense of the Academy, "even under the old regime the sciences had to a certain extent been organized into a republic and a kind of deference had protected them from the intrusions of despotism."[21] Of course in science, as in politics, a commitment to republican forms of civility in no way implied advocacy of egalitarian democracy.

Elections and promotions within the Academy called for delicate negotiations. The rules adopted in 1699 stipulated that, when a vacancy occurred and an election was to be held, the section concerned was to draw up a list of several candidates. All pensioners and honorary members were then entitled to vote in secret to determine which two or three candidates' names would be forwarded to the king. The final selection was made by the king, after being advised by his ministers. This procedure offered many opportunities for intense lobbying, yet it also made it difficult to ignore completely questions of scientific competence and the candidates' potential for achievement. Thus when Lavoisier turned his thoughts to the Academy of Sciences in the summer of 1764, he knew he faced a relatively open and rule-governed competition.

We have only a few tantalizing hints as to how Lavoisier deployed the forces marshalled by his family, mentors, and friends when campaigning for election to the Academy, but we do know he was not bashful in this regard.[22] We know considerably more about the investigative and rhetorical strategies he employed when seeking to convince the members of the chemistry section that he possessed the kinds of talent and outlook they were seeking in candidates for appointment. During his period of candidacy Lavoisier concentrated on two areas of research, one being the practical problem of lighting city streets, the other being the more theory-laden problem of analyzing minerals and mineral waters.

Sartine, the royal minister in charge of public order (*police*), was eager to provide Paris with well-lit streets. Initial experiments with different kinds of candles were begun late in the summer of 1764 and continued until the following May.[23] At the same time the Academy of Sciences, at Sartine's suggestion, announced a prize competition on how best to illuminate city streets. Lavoisier, eager to demonstrate that he was as concerned with the public good as he was with the advancement of chemistry, immediately attacked this problem on a number of fronts, his strategy being to address the issue comprehensively. No suitable proposals having been received by the first deadline in 1765, the prize was doubled to 2,000 livres and the competition was kept open for another year.

The essay Lavoisier submitted to the Academy, "On various ways of illuminating large cities," not only contains abundant evidence of his ambition and skill as an experimenter, it also reveals how completely he had internalized the ideology of state service. Like Nollet, Lavoisier believed there is a close connection between individual competitiveness and collective progress. "When minds, warmed by ambition and competition, are animated by ideas, a subtle fermentation excites them. In their common effort to win the prize they release new forces. In seeking to rise above others, they reach beyond themselves."[24] Sartine deserves credit, Lavoisier wrote, for harnessing these social and psychological forces to the needs of the nation.

> He calls to all citizens and invites learned men of all nations to send forward their suggestions. One of Europe's most esteemed tribunals will judge them. What a wonderful motivation for citizens! Inspired by such an appeal, how can they resist the enthusiasm arising from patriotic zeal? Even if they do not succeed, will it not be compensation enough to see their names mingled with those of others responding with zeal to serve the state?[25]

Lavoisier's enthusiasm had additionally been fired by the challenge of providing "calculations, physical and chemical experiments, and a theory applied to practice." Yet his sole concern was "to compete for the good of his fellow citizens." It was only following this breathless declaration of his own high-mindedness that he turned to the substance of the problem itself.

Lavoisier had not misjudged his audience and his prize essay made a good impression. The Academy subsequently devised a politically acceptable way to partition the available awards among those it wished to honor.[26] The essays received were divided into two classes. Those in the first were characterized as being "full of physical and mathematical discussions that evaluate the advantages and disadvantages of various useful means" for lighting city streets. Those in the second class were more practical and examined ways of lighting city streets that were already being used. The prize money was divided equally among the authors of the three best essays in the second class. Lavoisier's essay, the best in the first class, was to be published by the Academy. Sartine also obtained royal permission to have a gold medal struck for Lavoisier; it was awarded to him by the president of the Academy at a public meeting on 9 April 1766. Clearly this bright young man had convinced the king and his ministers of his commitment to applying science to the practical needs of the nation.[27]

Lavoisier employed a different but equally successful strategy to convince the members of the chemistry class that he was also committed to investigating theoretical problems of great concern to them. By November 1764, a few months after beginning his research on street lighting, he was also hard at work on the analysis of the mineral gypsum.[28] He had probably come upon this subject while working with Guettard, but the thrust of his research was much more theoretical than their mineral survey required. Clearly Lavoisier appreciated that when presenting himself as a candidate to the working members of the Academy, he had to treat theory rather than utility as his foremost concern. He realized, that is to say, that distinction in the republic of science was gained through the advancement of knowledge rather than its application.

This privileging of the theoretical over the practical was not unique to science, for academicians in all fields jealously guarded their monopoly over the articulation of theories and the adjudication of interpretive disputes. Academic hegemony concerning questions of method, interpretation, meaning, truth, and value was fundamental to the authority of academic culture.[29] All academicians would have rejected with contempt the suggestion that their status and role in French society depended primarily on the political, economic, or technical utility of their knowledge. They considered their high social status and the privileges associated with it nothing more than well-deserved public acknowledgement of their cognitive preeminence. Others might serve their king and nation by addressing problems of a practical nature, and such labor was certainly necessary and useful, but only an academician could speak definitively on the more elevated questions of truth and beauty.

Lavoisier provided evidence of his interest in distinctly chemical problems in the two papers on the analysis of gypsum he read to the Academy, the first on 27 February 1765, the second on 16 April 1766.[30] In a brief summary he said he considered the research described in these papers merely the first part of "an extensive investigation of all the minerals primarily by means of 'the wet way,' which is different from the technique of analysis employed by Pott [i.e., with fire or by 'the dry way']."[31] This was a bold yet carefully contrived claim. In committing himself to solvent analysis rather than destructive analysis by heat, Lavoisier was declaring his allegiance to a distinctive research program that had long occupied chemists in the Academy. As we shall see in chapter 4, it was also a research program that his master Rouelle had pursued. In their official report on the first of these papers Duhamel du Monceau and Jussieu noted and were impressed by the larger significance of Lavoisier's study of gypsum. Their recommendation led to his paper being published in the proceedings of the Academy.

In the spring of 1766 Pierre Joseph Macquer was promoted to the rank of associate, thereby creating an adjunct vacancy in the chemistry section. Lavoisier and his allies responded promptly. Evidence of his scientific promise and of his concern for the public good had already been presented; he read his second memoir on gypsum in March and received the medal for his essay on street lighting in early April. Shortly thereafter Lavoisier drafted two letters, one addressed to the president of the Academy and the other to its permanent secretary, proposing that a new section for experimental physics be created.[32]

These are curious documents and it is not known whether Lavoisier drafted them for his own use or for someone else, or indeed if they were ever sent. Their forcefully stated arguments provide further evidence of Lavoisier's early commitment to the methods of experimental physics. Yet it does seem rather odd that a young man who had as yet made no mark in science and was about to put himself forward as a candidate for election to the most select of all academies of science should propose that an entire new section be created.

The underlying purpose of these letters can perhaps be revealed by comparing the list of candidates with whom Lavoisier was competing and his suggested distribution of appointments in an expanded and reorganized Academy. The candidates for Macquer's position were identified on 23 April; the list included Cadet, Baumé, Lavoisier, Jars, de Mancy, Sage, Valmont de Bomare, and Monnet.[33] Before the senior academicians selected their nominees, Lavoisier drafted a note describing how current and future members of the Academy might be reappointed among the various sections in a reorganized Academy. It is noteworthy that Lavoisier assigned places in one or other of his seven sections to all the candidates for Macquer's position except Valmont de Bomare and Monnet. Since Lavoisier listed himself as an adjunct in chemistry, he evidently preferred an appointment in the chemistry section to one in the proposed section of experimental physics. But if he were able to arrange appointments for most of his competitors in one or other of the Academy's sections, then the likelihood of his obtaining an appointment in chemistry would be considerably increased. Such may have been his strategy when he drafted his two letters, but nothing came of it. The Academy nominated Cadet and Jars and the king appointed Cadet to the vacant adjunct position in chemistry.

As soon as the election was over Lavoisier threw himself whole heartedly into the mineralogical atlas project, which had recently been given official sanction. He embarked on a series of extended trips that kept him away from Paris for most of the next eighteen months.[34] In the summer of 1767 pre-election maneuvering began again, however, when Lavoisier's father sent him a report on what was in fact a continuing campaign. "I have seen Mr Miraldi [a pensioner in astronomy] several times," Lavoisier's father wrote.

40

> It appears that you are spoken of occasionally in the Academy and that nearly everyone has such a good opinion of you that Mr Maraldi thinks you will be nominated. In any case he evidently has decided that if it appears things are not going your way the elections will be postponed until you return.[35]

Lavoisier's father had also been seeing a good deal of Duhamel du Monceau, who had originally been elected to the Academy as a chemist and had subsequently been made a pensioner in botany. A few days later he wrote to his son again.

> Yesterday I saw Mr Duhamel, who sends you his best wishes. He regrets that you were not admitted to the Academy in the last election and says that since he voted for you, it is not his fault. He assumes there was considerable competition over who most deserved to be selected. Mr Cadet was the most fortunate. If nothing untoward occurs before your return, perhaps you could present a few memoirs, such as your essay on the illumination of cities. I think you have good reason to be hopeful.[36]

Lavoisier was back in Paris during the early months of 1768 and as busy as ever. His grandmother Punctis had died in January and Lavoisier received a considerable legacy from his mother's family. His father had released him from his legal status as a minor, which in France at that time extended to age twenty-five, and Lavoisier had begun considering how best to invest this legacy. On the advice of a family friend he purchased a share in the royal Tax Farm in the spring of 1768; his activities as a tax farmer will be discussed in greater detail in chapter 5. It appears that by 1768 Lavoisier had concluded that support for his election to the Academy was so strong that his best strategy was to act as if his admission was a foregone conclusion.[37] With the encouragement of friends and allies he proceeded calmly with plans to present to the Academy a series of papers based on his recent field research with Guettard.[38]

He read the first of these papers, "Research on the most precise and convenient techniques for determining the specific weight of liquids, whether for physics or commerce," on 23 March 1768.[39] The research program he followed was once again firmly anchored in one of the Academy's well-developed experimental traditions. His second memoir, "The character of the waters in parts of Franche-Comté, Alsace, Lorraine, Champagne, Brie and Valois," was read in two parts, the first on 4 May 1768 and the second three days later.[40]

Lavoisier began this latter memoir, as he did his essay on illuminating cities, with a ringing declaration of his determination to serve the entire nation.

> If society is interested in knowing about the waters of curative springs, whose astonishing effects have so often been celebrated in medical oratory, it is no less important to know about those waters utilized on a daily basis to satisfy the needs of life. It is these, in fact, on which the health and strength of the citizens depend. For while the waters of spas occasionally recall to life some of the more precious heads of state, it is the latter waters that sustain a much greater number on a daily basis by continually maintaining the order and balance of the animal economy. The examination of mineral waters is only of interest to a small and languishing part of society. The examination of common waters is of interest to the entire society and above all that active part whose arms are both the strength and the wealth of the state.[41]

In emphasizing the importance of public water supplies Lavoisier was again grasping hold of an issue that was already being intensely discussed in Paris. In 1753 the government had directed the chemists G. F. Venel and Pierre Bayen to prepare a report on all French mineral waters.[42] This effort proved fruitless, however, and Lavoisier may have hoped to revive and expand it by making it part of the more comprehensive survey he was conducting with Guettard. The academician Déparcieux had also campaigned for years both in the Academy and in the salons of Paris for the building of an aqueduct that would bring water from the Yvette river to the capital. Voltaire considered Déparcieux an admirable "citizen-philosopher" and in the summer of 1767 wrote to him from Ferney, saying that his project "would be worthy of the Romans, although unhappily we are not Romans."[43] Following Déparcieux's death in 1768 Lavoisier became the Academy's chief advocate of his scheme. In subsequent years he frequently returned to the problem of providing Paris with an adequate supply of clean water.

After announcing his concern for the public good, Lavoisier hastened to add that he was equally committed to the pursuit of scientific knowledge. In his paper on the waters of Franche-Comté he suggested that those engaged in field work transcend the dichotomy between the pursuit of knowledge that is socially useful and the search for knowledge that is scientifically valuable. While traveling with Guettard, he reported,

> we neglected nothing that might be of benefit to society or to the advancement of science, for we were convinced that no task is too difficult, no subject too extensive, – and this includes all the sciences, the arts and natural history – for scientists [*physiciens*] to encompass, in so far as is possible, while engaged in field research.[44]

Lavoisier did not, however, believe that the distinction between these two kinds of knowledge could be eliminated: "Beyond the social advantages realized through the investigation of waters and their importance to the health of individuals, this investigation has innumerable other advantages with regard to the progress of knowledge."[45] And it was in fact to the mineralogical and chemical aspects of the subject that he turned in the remainder of his memoir.

Lavoisier's campaign was brought to a successful conclusion during the latter weeks of May 1768. The adjunct chemist, Théodore Baron, had died early in March and an election was to be held soon thereafter. On 11 May, shortly after Lavoisier finished reading his long memoir on the analysis of waters, Nollet and Macquer reported favorably on its first part; a week later they did the same on the second part and recommended that the Academy publish the entire work.[46] On 14 May Lavoisier protected his proprietary interest in information he did not wish to make public at that time by having the secretary of the Academy seal and hold in his files a brief paper he had drafted on how to produce blue and yellow colors in fireworks.[47] A few days later the election was held and Lavoisier and Jars were nominated for appointment. The king, when choosing between them, gave special consideration to Jars's greater age and distinguished record of state service, even though Lavoisier had received slightly more votes.[48] But Lavoisier and his friends were not turned away empty-handed, for he too was appointed, as a supernumerary adjunct in the chemistry section. On 1 June 1768 Jars and Lavoisier were formally installed as members of the Academy. Lavoisier had not yet celebrated his twenty-fifth birthday, yet he had, by force of will and skillful maneuvering more than by actual service or scientific achievement, gained a place for himself among the tribunes of French science.

Those familiar with Lavoisier's later achievements cannot help but applaud the Academy's prescience in admitting him to membership at an early age. At the time, however, some of those who knew the young man personally were stunned by the commitment he had undertaken. In late May 1768 Augez de La Voye wrote to his cousin to congratulate him on his success. Partly in jest and partly in awe he wished Lavoisier well as he bid adieu to the pleasures of youth and embarked on the serious business of public life.

> I have long expected, my dear friend, to someday see your literary efforts honored by an academic appointment, but I did not expect it to happen so soon. In truth, dear friend, other young men will no longer want to consider you one of their colleagues; you have spoiled their game, for at an age when others seek only pleasure, frivolity, and dissipation, you work so seriously that you are accepted as a member of the Academy of Sciences in Paris. Nonetheless, I congratulate you with all sincerity for an appointment that would be an honor at any age, but especially at your age. Be careful, however, to take good care of your health while you are working, for you seem to me to be inclined to work a great deal.[49]

3

Experimental Physics

Lavoisier frequently commended the research methods of experimental physics, yet some modern commentators consider his statements on method little more than rhetorical glosses added to give the appearance of novelty to investigative practices that were in fact quite commonplace. It is indisputable, of course, that creativity in science involves much more than simply applying a general method to particular problems, and it would certainly be overstating the case to credit Lavoisier's innovative theories entirely to the methods he employed when investigating the fixation and release of gases in solids and liquids. But one does not have to choose between saying that a scientist's methodological convictions explain all of his or her achievements or that they are irrelevant to those achievements. Lavoisier did not argue that chemistry could or should become merely a branch of experimental physics, yet again and again throughout his scientific career he insisted that experimental physics deeply informed the way he fashioned, interpreted, and presented his chemical investigations. To make sense of these claims we need not puzzle over the general questions of the functions of methodology and experiment in science, as do philosophers of science. But the biographer cannot avoid asking in what way Lavoisier's frequently declared fidelity to the method of experiment informed his scientific beliefs and practices. What did these statements of principle mean to Lavoisier and how did they help him situate his research program and his theoretical achievements in the complex and contested world of eighteenth-century science?

Let us begin with a statement he made near the end of his life. By the middle of 1789 Lavoisier's oxygen theory of combustion had largely displaced the older phlogiston theory. Advocates of phlogiston, a hypothetical "matter of fire" that supposedly entered into chemical combinations but had not been isolated, explained combustion and its related phenomena as consequences of the release of phlogiston, whereas Lavoisier saw these phenomena as caused by the fixation of oxygen. A few unresolved problems remained and certain prominent chemists, most notably Joseph Priestley in England, were not prepared to abandon phlogiston entirely, yet by 1789 Lavoisier's new theories had for the most part triumphed. The new chemical nomenclature devised by Lavoisier and his colleagues, a language that incorporated his theory of oxidation, was being widely adopted, and the recently issued treatise Lavoisier wrote to introduce chemists to the new way of thinking was selling briskly. The chemical revolution was thus being brought to a close just as the great political revolution was beginning.

Although his new theories were hardly being neglected, Lavoisier never ceased worrying over how best to promote their wide acceptance. His task, as he saw it, was to convince young men who were just beginning their scientific careers that his system of chemistry was rational, natural and compelling.[1] He was particularly eager to capture the allegiance of those who had not yet committed themselves to any particular scientific doctrines. To do this, he repeatedly emphasized the methodological foundations of his new conception of chemistry. Indeed, in his retrospective accounts of the chemical revolution Lavoisier frequently treated method as the key to all his discoveries and innovations.

Consider, for instance, the recollections Lavoisier set down in a manuscript titled "The best way to teach chemistry" that he drafted sometime between 1790 and 1792.

> When I first took up the study of chemistry I was surprised by the number of difficulties that surrounded the approach to this subject, this despite the fact that my instructor taught clearly, was well-disposed towards students, and made every effort to help us understand. I had

46

had a good philosophy course and had taken the experimental course taught by the abbé Nollet. I had also grappled with some success with elementary mathematics as presented in the works of the abbé Lacaille and had taken his course for a year. I had in addition become familiar with the rigor with which mathematicians reason in their treatises. They never prove a proposition unless the preceding step has been made clear [*découverte*]. Everything is tied together, everything is connected, from the definition of a point to a line and on to the most sublime truths of transcendental geometry.

In chemistry everything was otherwise. From the outset one began by supposing rather than proving. I was presented with terms that were not defined and could not be defined without invoking knowledge that was utterly foreign and that I could not have known unless I had studied all of chemistry. When my instruction in this subject began, it was assumed that I already knew it.

When he had completed his first chemistry course with La Planche, Lavoisier recalled,

I wished to make an inventory of the chemical knowledge I had acquired. I realized I had an adequate grasp of all aspects of the composition of neutral salts and the preparation of mineral acids, the only chemical subjects for which there was exact and positive (*positive*) knowledge, but I had only obscure ideas about the rest of the science.

More confusion awaited him in the three courses he subsequently took with Rouelle. "The celebrated professor united a highly methodical presentation of his ideas with great obscurity in the way he expressed them." In the end, Lavoisier wrote,

I managed to gain a clear and precise idea of the state that chemistry had arrived at by that time. Yet it was nonetheless true that I had spent four years studying a science that was founded on only a few facts, that this science was composed of absolutely incoherent ideas and unproven suppositions, that it had no method of instruction, and that it was untouched by the logic of science. It was at this point that I realized I would have to begin the study of chemistry all over again.[2]

The didactic intention of Lavoisier's comments on method here, as elsewhere, is unmistakable. Yet the way he deployed these rhetorical resources toward the end of his life does not warrant the conclusion that he considered his methodological principles nothing more than devices for converting students to his new view of chemistry. Indeed, our initial response to these claims should be more sympathetic, for it may well be that Lavoisier had all along really believed his success as a scientist owed more to his methodological convictions than to the discovery of novel facts about nature. Today we find this view implausible largely because historians of chemistry have traditionally treated experimental discoveries, and most notably the deluge of new facts about the gases of the atmosphere that appeared in the 1760s and 1770s, as the primary engines of change in the chemical revolution.

This interpretive privileging of factual discovery over methodological critique has in the past been sustained by the assumption that, following the scientific revolution of the seventeenth century, all scientists were in essential agreement on how facts ought to be gathered, deciphered, and weighed. But as historians of experiment and of eighteenth-century science have amply demonstrated, the belief that after Newton all progressive scientists followed a common set of methodological practices is no longer credible.[3] Thus we should be prepared to take seriously Lavoisier's claim that his success owed at least as much to his utilization of a distinctive methodology as it did to the discovery of a novel set of chemical facts.

Lavoisier wrote out the recollection just quoted long after the period described. Of course, when composing this account he might have purposefully reconstructed what actually happened so as to validate his subsequent transformation of chemistry's theories and language. But there is considerable evidence that while still a young man Lavoisier experienced a Cartesian moment of illumination, a searing insight which revealed that much of what passed as received knowledge in his chosen field of chemistry was asserted rather than demonstrated, and that by adhering to a more rigorous method he could set chemical theory on firm foundations. As a result of this experience he decided in the mid-1760s that very little of the chemistry he had been taught was coherent or well-grounded and that, if chemistry was to advance, it had to commit itself to the methodological standards of mathematics and experimental physics. If this is so, then the revolution he effected in chemistry really did begin with a critical methodological assessment of the science as he found it. The transformation of chemistry he set out to effect remained little more than a programmatic proposal until, many years later, he reaped the rewards of his brilliant conceptual and experimental investigations of the role of air in combustion and calcination.

Lavoisier's attachment to the methods of experimental physics can be traced back to the earliest years of his career. In 1764, when only twenty-one years old, he drafted the first of what became a series of outlines for an introductory chemistry course. Marco Berreta has edited the manuscript copy of this earliest proposal and has demonstrated both that Lavoisier's first draft of a chemistry course incorporated many of the concepts, procedures, and subjects normally treated in courses on experimental physics and that it was unlike any of the chemistry courses then being offered in Paris.[4] While Lavoisier evidently did not wish to duplicate the course in experimental physics taught by his master Nollet, he clearly believed that a similar course in chemistry would find an audience. The course he outlined incorporates the methodological principles that were fundamental to Nollet's lectures and clearly challenged the courses taught by apothecaries such as La Planche and Rouelle. And, as will be explained more fully later in this chapter, it also posed a challenge to the more comprehensive Newtonian approach to chemistry championed by Buffon. This early proposal therefore tells us a great deal about how Lavoisier chose to position himself as a chemist during the year in which he completed his legal studies and turned his attention to the Academy of Sciences. What is most striking in hindsight is how powerfully the methodological orientation he adopted in 1764 continued to guide his scientific thinking throughout the remaining three decades of his life.

The assiduous use of precise instruments was the central tenet of experimental physics as it was practiced in the 1760s, and it was this commitment that Lavoisier made central to his proposed introduction to chemistry. Investigative instruments made analytic quantification possible, and experimental physicists believed that precise quantification would eventually lead to mathematical rigor. This program of research was more than a Cartesian dream, for during the middle decades of the century more refined instruments were being constructed and measurements of ever greater precision were being made with them. Lacaille and Nollet, Lavoisier's mentors in experimental physics, were both noted instrument-makers and Lavoisier had had the good fortune to study with them at precisely the historical moment when experimental physics was emerging in France as the pre-eminent approach to the study of nature.[5] For Lavoisier it was a fortuitous conjunction that shaped his entire career and achievement.

The construction of exact instruments and how they should be used to obtain precise quantitative data are central concerns in Lavoisier's 1764 manuscript.[6] Seven topics are addressed in his outline: (1) the definition of chemistry, (2) the principles of bodies, (3) fire as a cause of dilation, (4) the pyrometer, an instrument for measuring the expansion of a metal bar when heated, (5) thermometers, (6) other properties of fire, and (7) electricity. Three features of this outline deserve special comment. The first is Lavoisier's commitment to constructing instruments, in this instance pyrometers and thermometers, which would enable him to measure the specific effects of heat on different substances. The second is his suggestion that fire, which he thought of as an extremely subtle fluid that surrounds and penetrates all bodies, appears in one of its modifications as light. This claim took on added significance when Lavoisier went on to say that Newton's elaborate theory of how light and colors are transmitted, while useful in physics, does not meet the needs of chemistry. Lavoisier suggested that colors are caused by waves like those that transmit sound, a theory that had previously been proposed by the mathematician Leonard Euler. Lavoisier did not pursue this suggestion with vigor, either in this text or in his later research, but his early search for alternatives to Newtonianism, while hedged by acknowledgements of the great man's achievements, is an indication of the extent to which he was determined to follow the Cartesian Nollet rather than the Newtonian Buffon when constructing theories to explain chemical phenomena.

Lavoisier's discussion of electricity, the third notable feature of his 1764 manuscript, is brief, but its very presence in a proposed chemistry course is unusual. He describes the investigations and theories of both Benjamin Franklin and Nollet, but avoids taking sides in the theoretical controversy stirred up by their partisans. His account of Franklin's famous kite experiment is particularly interesting, for it shows that, rather than simply seeking to improve chemical theories by marrying them to the facts and concepts of physics, Lavoisier was most impressed by the methodological rigor of experimental physics, its ability to demonstrate that its claims are true. He wrote that the experiment in which Franklin used a kite and a key to capture lightning and store it in a Leyden Jar raised the identification of electricity and lightning from an analogy to a demonstration. It was this experiment's ability to provide demonstrative evidence, rather than its ability to reveal surprising facts, that Lavoisier valued most highly. Although investigating the connections between electricity and chemistry did not prove to be a particularly rewarding line of research for Lavoisier, the very fact that he explored this connection in the 1760s indicates how thoroughly he was committed to using the resources of experimental physics in his attempt to reconstruct the science of chemistry.

Two years later Lavoisier made his commitment to experimental physics the centerpiece of an institutional proposal. In a letter to the secretary of the Academy that he drafted in April 1766 he argued that experimental physics should be given a central role in the Academy.

When the Academy was established in 1666, it was organized into two classes, one for physics and the other for geometry. Since then, experimental physics has escaped from the shadowy laboratories of earlier chemists and, guided by the sure hands of Huygens, Mariotte, and Perrault, has begun to take on a new form. Firmly based on experiments and facts, it has steadily advanced. It has called into question systematic philosophy and its progress has been so rapid that it already constitutes a very considerable branch of science. Why then was this science completely ignored when the Academy was reorganized in 1699? We must conclude, no doubt, that the subjects it studies were not numerous enough to support the creation of a separate section. That is how the progress of experimental physics was brought to a halt in France and foreigners benefited from what we cast off. It is obvious that appointments, more than anything else, promote the development of new fields; they ignite in young spirits the animating fire of competition that urges them on to great achievements.

Since that time several prominent individuals whose works have contributed to the glory of this Academy have given new life to experimental physics. It has been infused with greater vigor than before, it has enriched the sciences and the arts, and, with the assistance of experiments, it has brought certitude to all aspects of our understanding.[7]

This letter is most revealing. According to Lavoisier, chemists were the first to make systematic use of experiments, yet the three academicians who first applied this method to other subjects were physicists. But this initiative was not adequately encouraged in France and leadership soon passed to investigators in other nations – no doubt Lavoisier was here thinking primarily of Great Britain and Holland.[8] Happily, experimental physics had enjoyed a revival in France and, more particularly, in the Academy – Lavoisier clearly had in mind the research of Nollet and his mentor Réaumur. This revival deserved support, for experimental physics brings "certitude to all branches of our understanding." This is the core belief, here stated concisely and directly at the beginning of his scientific career, that gave Lavoisier confidence as he set about transforming into a proper science the burgeoning but ill-grounded subject of chemistry.

Lavoisier, like his teacher Nollet, believed that experimental physics could carry scientific investigators past the two great hazards of credulity and skepticism. The systems of natural philosophy proposed by Descartes, Newton, and Leibniz begin with accounts of certain properties of matter which are assumed to be of fundamental importance and proceed largely by drawing out the implications of those assumptions. Experimental philosophy seeks to replace these excessively rationalistic systems with bodies of knowledge consisting of inductive generalizations that are confirmed at every step by experimental demonstration. Experimental philosophy vanquishes skepticism, the claim that we cannot acquire any reliable knowledge about nature, by demonstrating that we can, if we rely on carefully designed and well-conducted experiments, construct limited but verifiable explanations of natural phenomena. It avoids the hazards of credulity or unfounded belief by insisting that reason can be relied upon only when it is disciplined at every step by experimental recourse to nature.

Lavoisier, like many of his contemporaries, was invigorated by this eighteenth-century version of positivism, which added a crucial middle term to the seventeenth-century distinction between facts and causal explanations. Lavoisier would not have disagreed with the academician Fontenelle, who wrote in 1699 that

> We are forced to look upon present-day science, at least physics, as if it were in its cradle. Hence the Academy is only at the stage of gathering an ample store of well-founded observations and facts, which will one day become the basis for a System. For Systematic Physics must refrain from

building its edifice until Experimental Physics is able to furnish it with the necessary materials.[9]

Yet Lavoisier would have found Fontenelle's advocacy of experimental physics methodologically inadequate, for it says nothing in detail about how one moves from an accumulation of accurately described facts to the construction of true theories. It was the method of reasoning by experiment more than the discovery of novelty by means of experiment that Lavoisier found most promising. Thus in his 1765 analysis of gypsum Lavoisier wrote that "before beginning the analysis of this mineral, I will describe several preliminary experiments that will serve as lemmas for the propositions that I will demonstrate."[10] Lavoisier looked forward to formulating theories having all the certainty associated with deductive and demonstrative rationality. His theories, however, unlike the propositions of geometry or the speculations of natural philosophy, would be verified at each step by appropriate and compelling experimental facts.

Lavoisier repeatedly invoked the standards of proof used in experimental physics when criticizing received chemical theories. The main purpose of the paper on the specific gravity of liquids that he read to the Academy of Sciences in March 1768 was to describe a constant immersion hydrometer which, he believed, would bring a new level of precision to the study of solutions. In a digression on the inadequacy of certain widely accepted chemical theories, Lavoisier summarized several explanations of how air is fixed in, that is to say combined with solids. He concluded, with a harshness that was rather daring on the part of a candidate for election to the chemistry section, that "when we put these questions to chemistry it replies with such useless names as resemblances, analogies, and frictions . . . which clarify nothing and only have the effect of training the mind to be satisfied with words."[11] When he wrote these dismissive comments on the theoretical standards of chemistry, Lavoisier had no better explanation for the fixation of air to offer. Yet he was convinced that if appropriate instruments were applied to problems on the interface between physics and chemistry, the facts revealed could be woven into a detailed theory that would simply blow away the unsubstantiated speculations most chemists were willing to accept as explanations. It was his faith in a method of investigation and theory construction, rather than possession of a well-developed alternative theory, that made him so dauntingly confident at that early stage of his career.

In August 1772 Lavoisier again found fault with one of the central concepts of chemistry, in this case the hypothetical matter of fire to which the German chemist G. F. Stahl had given the name phlogiston. As before, he did not offer an alternative theory of heat and fire, yet that did not prevent him from characterizing the concept of phlogiston as imprecise and unsubstantiated. He insisted, in a memorandum not intended for publication, that although phlogiston was widely invoked to explain changes caused by intense heating, the theory lacked conviction. "One must admit that we still have so little understanding of the nature of what we call phlogiston that we cannot say anything precise about it."[12] A few months later, probably in November 1772, Lavoisier began drafting a memoir on the weight gain that occurs when metals are calcined (i.e. roasted), and once again he voiced reservations about Stahl's phlogiston theory.

> First of all, it is clear that Stahl's theory on the calcination and reduction of metals is extremely faulty and must be modified. It asserts that all calcinations only involve a loss of phlogiston, even though it has been proven that there is an absorption of air at the same time there is a loss of phlogiston.[13]

Experimental physics showed Lavoisier both how to clear away the underbrush of unsubstantiated speculation and how to go about setting chemistry on firmer foundations. To formulate a more adequate theory of the weight gained in the calcination (i.e. oxidation) of metals, he had to find facts capable of transforming his hypotheses into demonstrably true theories. Facts, he noted, drive science onwards, sometimes slowly and incrementally and sometimes in a revolutionary tumult. Just how they do this was a question Lavoisier wrestled with while struggling to grasp the significance of his early experiments on combustion and calcination.

> Among the facts that chemists discover are many that have little significance until they are connected together and assembled into a system. Chemists may withhold facts of this sort and ponder in the quiet of their

studies how to connect them to other facts, announcing them to the public only when they have been pulled together so far as is possible.

There are other types of experiments that overthrow established systems of thought and open new lines of experimentation and reasoning, which in a word appear capable of effecting a revolution in the science. Facts of this sort should be reported to the public immediately. To withhold them from the public is to retard the progress of the art; it is in effect to conceal something from the community of the learned; it is above all to go against the wishes of the academic institution of which we are all a part.[14]

Lavoisier viewed chemistry as a domain of problems, concepts, and theories. Although he wrote out several preliminary drafts for chemistry courses, he never completed a comprehensive account of the science, as did Rouelle in his public lectures, Macquer in his *Dictionnaire de chymie*, and, at a later date, Fourcroy in his textbooks. He was drawn instead to problems that arose on the borderland between chemistry and physics, where he found challenges that appeared amenable to the techniques of investigation and patterns of explanation he found persuasive. He was, beyond question, a well-trained and competent chemist, but while chemistry was Lavoisier's chosen field, as a theorist he followed the methods of experimental physics. It is therefore anachronistic and unhelpful to try to categorize him as either a chemist or a physicist. He roamed across both subjects while seeking to construct unassailable theories that would contribute to the advancement of both sciences.[15]

Lavoisier was not alone in seeing great promise in the methods of experimental physics, but among chemists he appeared exceptionally single-minded in his commitment to this approach. In 1766 P. J. Macquer noted in his popular *Dictionary of Chemistry* that happily "chemistry was beginning to be cultivated according to the methods of well-grounded physics."[16] Two years later Macquer, along with Nollet, was asked to evaluate the two papers on the analysis of waters that Lavoisier read to the Academy in the early months of 1768. His first paper, they reported, was "by a man who knows a great deal about physics and who employs that knowledge with great wisdom." In his second paper, they continued, Lavoisier "astonished all physicists" by determining the temperature at which water reaches its maximum density, yet he persuaded them by "performing in the presence of the Academy the experiments that support this important discovery."[17] Lavoisier evidently took pains to present himself to the Academy as an advocate of the methods of experimental physics and clearly that is how he was perceived.[18]

Lavoisier's most famous statement of his scientific intentions is contained in a research memorandum he wrote out for his own edification in the first pages of a new laboratory notebook he began using in February 1773.[19] Convinced that his initial experiments on combustion, calcination, and the fixation of air had opened up a promising avenue of investigation, he outlined a research program that would, he believed, enable him to formulate new concepts and theories of revolutionary importance. His expectations proved to be prescient, although at the time he had no idea how he would solve the many detailed problems that were bound to arise along the way. His first task, he realized, was to master the relevant literature. He noted in his laboratory notebook that "the importance of this subject encouraged me to review all these studies, which appeared to me destined to bring about a revolution in physics and chemistry." By the end of the year he had completed a survey of earlier work and had prepared reports of his own experiments. These materials were then published in his first book, which appeared in 1774, the *Physical and Chemical Essays* (*Opuscules physiques et chimiques*).[20]

Lavoisier originally planned to publish additional collections of *Physical and Chemical Essays*, but he was distracted by other endeavors and no further volumes were sent to press.[21] Some eighteen years later, however, he again turned to the task of editing his now numerous memoirs and papers. And at the end of his career, as at its beginning, the title he selected to describe his life's work mentions both chemistry and physics. But this project too was disrupted by events and the first two volumes of Lavoisier's *Memoirs on Physics and Chemistry* (*Mémoires de physique et de chimie*), edited by his widow, were not published until a decade after his execution.[22]

Experimental physics was flourishing in France at the beginning of the 1760s. It was, however, only one of several distinct approaches to physics that continued to compete for hegemony.[23] An older view of the subject, one that most working scientists in France no longer found rewarding, conceived of physics as Aristotle had defined it, that is, as a comprehensive philosophy of nature. A more lively tradition expanded on the mathematical realism of the seventeenth century and sought to express natural laws, as Newton had expressed the law of universal gravitation, in mathematical terms. Although mathematical physics was still being actively pursued in the latter half of the century, few scientists continued to believe that mathematization was the only credible warrant for scientific truth. And shortly after Nollet's installation as the first French professor of experimental physics, Denis Diderot boldly if incorrectly ventured the opinion that the age of the mathematicians was at an end.

> We find ourselves in the midst of a great revolution in the sciences. The ways in which ethics (*morale*), literature, natural history and experimental physics are being studied today convinces me that before another hundred years have passed, one will not be able to find three distinguished mathematicians (*géomètres*) in Europe. This science will come to an abrupt halt and remain where the Bernollis, the Eulers, the Maupertuises, the Clairauts, the Fontaines and the d'Alemberts left it . . . In the years to come their works will stand like the pyramids of Egypt, the combination of hugh masses and dense hieroglyphs inspiring in us a terrifying sense of the power and resources of the men who built them.[24]

Yet another approach to physics looked primarily to seventeenth-century theories of matter and sought to use them to explain a broad range of observable phenomena. The contending schools of Cartesians, Newtonians, and Leibnizians championed incompatible assumptions about the nature of elementary matter and their partisans quarreled fiercely in the salons, academies, and journals of learned Europe. And finally, but not exhaustively, there were the experimental physicists, whose epistemological claims, when compared to those of their more vaunting contemporaries, appeared surprisingly restrained.

Experimental physics itself consisted of a complex set of research programs that encompassed several specific sub-traditions. These included distinctive approaches to such activities as the writing and publishing of introductory textbooks, the utilization of instruments in investigative practices, the interpretation of factual evidence, the development and appraisal of theories, and the cultivation of master–pupil affiliations.[25] Experimental physics' common distinguishing feature was its methodological insistence that what we claim to know about nature must be founded on and verified by demonstrable facts. Experimental physicists respected and made use of reason and mathematics, but they did not trust rational explanations unless they were supported by physical experiments designed to confirm their veracity.

Experimental physicists were not the only scientists who attributed great importance to the role of facts in science, but they were distinctive in insisting that the ultimate warrant for truth claims in science is provided by facts obtained from purposeful experiments. Experimental physics was more positivistic, in the eighteenth-century sense of this term, than other approaches to natural philosophy in being more concerned with the fit between theory and observable reality than with the logical integrity of comprehensive theories. Of course experimental physicists sought to minimize inter-theoretic incoherence – it was not likely that a chemical theory of elective attractions that contradicted the theory of gravitational attraction would find a large following. Yet rather than resolving such problems by adding ad hoc hypotheses to their fundamental assumptions, experimental physicists normally sought to avoid problems of this sort by applying their concepts and theories only to those phenomena they were capable of explaining. Philosophically inclined scientists like Buffon considered experimental physics timid and narrow; those who, like Lavoisier, were more concerned with credibility and certainty saw it as sure-footed and incremental.

In eighteenth-century France experimental physics, like natural history, was largely a creature of polite society. It flourished in the lecture halls of Paris, to which hundreds of leisured amateurs flocked year after year. In such an intensely self-conscious urban culture, where many varieties of science were offered to a demanding and actively involved public, personal preference was an important factor when choosing a methodology. Those who wished to avoid the distractions of public controversy tended to shun the more speculative philosophical sects and the often sharp, although always formally courteous, quarrels they spawned. It therefore seems reasonable to suppose that

Lavoisier found the methods of experimental physics attractive because, while they were centrally concerned with theory, they gave rise to a relatively restrained and polite form of scientific discourse. These features looked very appealing to a man who preferred to operate from a position of consensual and institutional authority and who was involved in many sensitive public undertakings besides science.

Those with a greater appetite for confrontation, manipulation, and adulation found the self-restraint of experimental physics bloodless. Eager to generalize and hungry for public approbation, they exhibited aristocratic or populist assertiveness rather than bourgeois caution. Vigorous advocates of conjectural hypothesizing and combatants who relished the cut-and-thrust of debate had little patience with those who wished to have all their facts in order when hazarding theoretical interpretations. One such controversialist, and one whose theories had a considerable impact on French chemistry in the eighteenth century, was Georges Louis Le Clerc, comte de Buffon. A brief look at his program for the advancement of chemistry will help to sharpen our appreciation of the methodological choices Lavoisier faced at the outset of his career.

Today Buffon is remembered as a natural historian and literary stylist, but at the beginning of his career he was known primarily as a mathematician and a Newtonian.[26] In 1735 he prefaced his translation of Stephen Hales's *Vegetable Staticks* with a ringing endorsement of experiment. Experiment was, he said,

> the method that my author followed, that of the great Newton, and the method that Bacon, Galileo, Boyle, and Stahl have adopted and recommended. It is also the method the Academy of Sciences made obligatory, and its illustrious members Huygens, Réaumur, Boerhaave and others have employed it and continue to employ it effectively. It is, in short, the way the great investigators (*les grands hommes*) have always followed and still follow.[27]

Buffon was twenty-eight years old when he penned this encomium to experiment and he had been a member of the Academy of Sciences for one year. It soon became evident, however, that his commitment to experiment was largely rhetorical and that he was in fact more impressed by Hales's use of the Newtonian theory of attraction than by his experiments.[28] Buffon, like Voltaire, took great pleasure in enlarging on the speculative possibilities of Newtonianism and he frequently employed the English doctrine as a stick to beat his French adversaries. He had little interest in the minute examination of nature and characteristically dismissed Réaumur's painstaking six-volume natural history of insects with the remark that "a fly ought not to occupy a greater place in the head of a naturalist than it does in nature."[29] In the oration he delivered in 1753 at his inauguration as a member of the Académie Française, Buffon declared that the value of a scholar's work lies not in the discovery of facts but in their organization and presentation.[30] When organizing and presenting his own comprehensive philosophy of nature, Buffon found the Newtonian concept of attraction especially useful. By the end of the 1730s Voltaire was calling him one of the most prominent Newtonians in France.[31]

Although a staunch Newtonian, Buffon was prepared to revise certain aspects of his master's doctrines. Unlike Newton, he rejected all notions of providential intervention and teleological explanation when reasoning about nature; he subscribed instead to the secular naturalism that was becoming commonplace in French science in the second half of the century. Buffon also insisted there is an unbridgeable chasm between the deductive certainty of mathematics and what can be known about the physical world, a view that his friend and disciple Diderot adopted as well for his own purposes.[32] Such a view would seem to strengthen the skeptical critique of science, but Buffon met this challenge by elaborating a probabilistic account of physical knowledge and never lost his faith in the possibility of gaining access to the fundamental laws of nature. Newton's law of gravitational attraction was Buffon's best evidence that his confidence in the power of the mind was not misplaced. Indeed, Buffon was so taken by Newton's achievement that he considered the concept of attraction the key to solving all the problems of natural history and chemistry.[33]

In 1739 Buffon was promoted to the rank of associate in the Academy and transferred to the botany section; in the same year he was made director of the King's Garden (Jardin du Roi), the zoo and natural history museum that became the present-day Muséum National d'Histoire Naturelle. Buffon's senior colleague Réaumur felt slighted when this prestigious position was awarded to the popular younger man; the appointment may have been the source of the bitter animosity between the two men that first arose at about this time.[34] This rivalry wore many masks as it matured into a full-blown feud. It not only touched upon the simmering antagonism between French Cartesians and Newtonians, but also on the competing styles of popular speculators, such as Buffon, and of more restrained observers and experimentalists, such as Réaumur. Quarrels of this sort routinely entangled all those associated with their principals. By the 1750s Nollet, who two decades earlier had spent several years as Réaumur's assistant, had been drawn into an acrimonious dispute over electrical theory stirred up by the partisans of Réaumur and Buffon.[35] Younger apprentices are keen and interested observers of the fissures of discord that so often lie just below the polite surface of science. There can be little doubt that by the time Lavoisier had completed his scientific education, he knew all about this smoldering battle between Nollet and Buffon.

René Antoine Réaumur died in 1757, when Lavoisier was only fourteen years old, but his reputation as a naturalist and as one of the Academy's foremost experimental physicists endured for years.[36] In 1713 the Academy had charged him with editing a vast industrial encyclopedia Colbert had urged it to undertake, and for the next few years Réaumur investigated the processes by which metals and porcelain were produced. He had also worked to improve the instruments used by physicists, especially the thermometer, for which he devised a new scale. A devoted academician, he contributed to many of its investigations and frequently held prominent administrative positions. Nollet considered Réaumur his mentor as well as a close friend and let it be known that in his opinion Réaumur was a model scientist. Given the factionalism that permeated French science, one can imagine what Lavoisier thought about the Buffonians who maligned the experimental physics of Réaumur and Nollet. But in public, and even in his private papers that have survived, Lavoisier was exceedingly cautious; he never openly presented himself as a partisan in this affair.[37]

Nollet entered the Academy in 1739, when he was appointed to the position that Buffon vacated following his promotion to associate. Nollet's courteous and agreeable manner no doubt helped forestall conflict during his first decade as a member, but in 1750 the recent success of his electrical theory made him an inviting target for Buffon and his supporters. The quarrel that ensued was thus a direct spinoff from the long-standing feud between Réaumur and Buffon. Its history reveals much about the contentiousness and complexity of eighteenth-century culture.[38]

In 1749 Buffon prefaced the first volume of his monumental *Natural History (Histoire naturelle)* with a preliminary discourse designed to clear the ground he intended to occupy and lay out the principles to be followed by all those who wished to understand nature. His chief targets were the taxonomic system of Linnaeus, which he dismissed as artificial, and ahistorical descriptions of nature of the sort provided by Réaumur.[39] A year later Réaumur's friends responded in kind in a five-volume set of *Letters to an American (Lettres à un amériquain)* that was published in Hamburg. While still smarting from this counterattack, Buffon came upon a copy of Benjamin Franklin's *Experiments and Observations on Electricity,* the first volume of which had recently been published in London. He quickly realized that Franklin's theory differed radically from that of Nollet and hence provided him with another opportunity to use a work written in English to pummel his opponents. Buffon promptly arranged to have Franklin's book translated into French. Neither Nollet nor his theories were mentioned in the preface, but everyone concerned was aware of the calculated insults directed at Nollet that Buffon sprinkled liberally throughout the text. Buffon reported delightedly that Nollet was "dying of chagrin from it all."[40] Nollet did not let the challenge pass unanswered, however, and the quarrel became increasingly highly charged.

Buffon had picked the right horse when he put his money on Franklin, for Franklin's theory, when refined to take account of Nollet's criticisms, provided better explanations for certain electrical phenomena. The controversy became rather technical and complex as it proceeded throughout the 1750s, and the Buffonians had difficulty keeping up with new developments. Nollet persisted in his rear-guard action until his death in 1770, but the tide was running against him. Younger men in other countries, who were just entering the field, were concentrating their attention on the recently invented Leyden Jar, a powerful electrical condenser, and Franklin's theory of the electrical fluid provided the best available account of its effects. In Paris contention between Nollet and the Buffonians confused and eventually paralyzed electrical research; by 1760 France was no longer providing leadership in the investigation of electrical phenomena.[41] Lavoisier's innate aversion to controversy was no doubt reinforced by this prolonged feud, one that damaged Nollet's reputation and effectively stalled French investigations into a most promising subject. Perhaps he was attracted to chemistry in part because it offered him a way to revive experimental physics in France without becoming involved in a draining controversy. He could instead focus on previously neglected problems on the interface between chemistry and physics, problems such as the fixation and release of the elements water, air, and fire.

If Lavoisier did indeed turn to chemistry partly to avoid controversies of the sort that had made electrical research a battlefield, he must have known that chemistry also had its contested ground. Buffon left his mark on this field too. He had, predictably, set himself up as an advocate of Newtonian chemistry, and the research program he championed flourished in the 1770s and 1780s.[42] These, of course, were precisely the decades in which Lavoisier was working out his own novel concepts and theories. But as historians of chemistry have repeatedly emphasized, Buffon's apodictic statements on how chemical theories ought to be constructed played no part in the revolution Lavoisier led. As Maurice Crosland has noted, "Buffon made very little direct contribution to chemistry, [and] it was sheer impudence for [him] to come along and say that the obscurity of chemistry was largely due to the fact that little attempt had been made to generalize its principles."[43] Yet it must also be remembered that in the mid-1760s Lavoisier had no way of knowing history would vindicate him, and the looming presence of Buffon must have given him many anxious moments.

In 1765 Buffon included as a preface to volume XIII of his *Natural History* a second programmatic statement he called his "second view."[44] One of its purposes was to insist yet again that Newton's theory of universal gravitation is fundamental to the explanation of all natural phenomena, including those examined by chemists. Buffon claimed that

> the laws of affinity by which the constituent parts of different substances are separated from one another so that they can recombine among them-

selves and form homogeneous bodies are the same as the general law by which all the heavenly bodies act on one another. They act equally and in the same manner with regard to masses and distances: a particle of water, of sand or of metal acts on another particle just as the globe of the earth acts on that of the moon.[45]

A mediating theory was clearly needed to connect the universal law of gravitation to the diverse phenomena of chemistry. Buffon provided it by declaring that the shapes of the constituent particles cause gravitation to act in the various ways observed by chemists.

> All matter is mutually attracted according to the inverse square of the distance and this general law can only be altered in specific attractions by the shape of the constituent particles of each substance, because this shape enters as an element in the distance.[46]

The argument is Newtonian but not Newton's, for, as Buffon carefully pointed out, Newton himself had not quite grasped how to apply his general theory to the phenomena of chemistry.

> Newton suspected that chemical affinities . . . act according to a law very like gravitation, but he seems not to have realized that all these particular laws are only simple modifications of the general law and that they only appear different because over the very small distances involved, the shape of the atoms counts for more than the mass and that the shape therefore matters greatly when determining distance.[47]

Buffon published this uncompromising restatement of his views in 1765, the year in which the academic astronomer A. C. Clairaut died. This was more than coincidence, for the "second view" was an unblushing declaration of victory in yet another of the quarrels into which Buffon had entered. This controversy began in the 1740s, when Clairaut and several other astronomers noted that their calculations of the Moon's motion led to results that differed from those reported by Newton.[48] Clairaut suggested that perhaps the force of attraction between the Earth and the Moon was not exactly in proportion to $1/r^2$. Buffon, ever the defender of orthodox Newtonianism, objected on grounds that were fundamentally metaphysical, and so the dispute began. It waxed and waned over the years as new evidence was considered, but the disputants were basically arguing past one another.

Buffon was characteristically adamant and, as luck or intuition would have it, had once again backed the right horse, for upon further examination it was demonstrated that the anomalies that had troubled Clairaut could be adequately accounted for with Newton's lunar theory.

Lavoisier probably considered Buffon's insistence on the centrality of attraction pure speculation, a hypothesis that could not be subjected to experimental test, but other chemists treated it with greater respect. In his 1766 *Dictionary of Chemistry* P. J. Macquer gave the Buffonian program much more prominence than he had accorded to Newtonian chemistry in his earlier three-volume *Elements of Theoretical and Practical Chemistry* (*Élémens de chymie théorique et pratique*), which first appeared between 1749 and 1751. In the *Dictionary* he asserted that chemical phenomena are without doubt caused by the reciprocal force of gravitation. This is why, he insisted, examining the weights of constituents "is the most important and crucial consideration for any general theory of chemistry."[49] Buffon was pleased to find his cause taken up by one of France's foremost chemists; in 1771 he arranged to have Macquer appointed to the chemistry professorship at the King's Garden.

Buffon's fellow Burgundian, Guyton de Morveau, also marched under the banner of Newtonian chemistry. The two men first met in 1762, two years before Guyton became interested in chemistry. His first book, *Academic Digressions* (*Digressions académiques*), was published in 1772. The first part is devoted to a lengthy discussion of phlogiston, the second to implementing Buffon's research program. Guyton hoped to relate chemical affinities to the shapes of constituent particles by studying solutions and crystals. Buffon initiated a correspondence shortly after this book appeared and Guyton was soon basking in the light of his approbation. In his three-volume *Elements of Theoretical and Practical Chemistry* (*Élémens de chymie, théorique et pratique*), published in 1777–8, Guyton hailed Buffon as "the Newton of France" and declared that "since Newton did not see clearly how chemical affinities operate, it was left for Mr Buffon to illuminate these matters for us."[50]

Lavoisier was less impressed. He was, of course, familiar with the chemical study of relative attractions and considered it a useful way to organize substances according to their distinctive chemical properties. Chemists had long known that certain substances tend to combine more readily with some substances than with others and many eighteenth-century chemists compiled tables that listed substances according to what were called their elective affinities. Thus, for instance, in the well-known table of affinities that E. F. Geoffroy presented to the Academy in 1718, the first column is headed "acids" and has listed under it, in the following order, fixed alkali, volatile alkali, absorbent earth, and metallic substances. Sulfur stands at the head of another column and has listed under it fixed alkali, iron, copper, lead, silver, antimony, mercury, and gold.[51] These series convey the information that the substance at the head of the column will combine more strongly with a substance close to it on the list than with one further away. Thus in principle, if copper is added to a compound composed of sulfur and mercury, the copper will displace the mercury in the compound.

Geoffroy, and later Lavoisier, realized that however useful affinity tables might be for taxonomic purposes, they cannot explain the different degrees of attraction among various substances. One was free to believe that the underlying cause is Newtonian attraction, as Buffon did, but the evidence provided by elective affinities did not compel one to do so. Geoffroy, in fact, was not a Newtonian. Lavoisier was therefore happy to treat elective affinities as part of the discourse of chemistry and to use the concept when describing his findings, but he did not subscribe to the Buffonian belief that Newtonian attraction is fundamental to all chemical action. On this issue, as on so many others, the speculative Buffon and the positivistic Lavoisier parted company.

Lavoisier was hardly being original in preferring the cautious progress of experimental physics to the unrestrained assertiveness of Buffonian natural philosophy. His methodological preferences embodied values he had acquired while studying with Nollet and, as we shall see, these values were also those of the eminent elder statesman of science, Jean d'Alembert. Those who wished to be experimental physicists had to have access to good laboratory instruments. Nollet, being a man of modest means, had been obliged to build his own apparatus. He was successful in this and was well-known as an instrument-maker and vendor years before he distinguished himself as a public lecturer. When Nollet spoke of experimentation he did so with an immediacy of experience that Lavoisier evidently found convincing.[52]

Nollet's lecture course began with a history of physics organized according to "the revolutions by which it has been established."[53] Descartes had proposed a new method that dramatically reformed the subject, yet Nollet wished to avoid debates over systems. With a subtle allusion to a well-known line in Voltaire's *Philosophical Letters* (*Lettres philosophiques*), he ridiculed those who encouraged speculative disputes by playing the Newtonian in Paris and the Cartesian in London. The physics Nollet offered was grounded in the senses and supported by facts. The subjects addressed were chosen for their interest, their novelty, and the likelihood that experiments would render them intelligible. His experiments were not for show alone; they were fundamental to his conception of physics.

> Observation and experiment are the most reliable, indeed the only, means by which an investigator can contribute to the progress of physics. In the first of these one spies on nature to discover her secrets; in the second one commits violence to force her to speak. But whichever of these means one uses, one must proceed in the prescribed manner. This difficult art requires a natural disposition and unusual attentiveness, attributes that are seldom easily acquired.

It was above all to the art of experiment and its consequences for physics that Nollet wished to introduce the students who attended the "school of physics" he had established in Paris.

Nollet's presentation of his subject cannot be dismissed as simply "physics for the ladies," the posturing of a popular showman who performed dazzling but theoretically meaningless experiments while avoiding the rigors of mathematics. No less a mathematician that Jean d'Alembert, one of the few *philosophes* whose scientific and literary achievements were equal to those of Buffon, publicly commended Nollet's experimental approach to physics. This endorsement can be found in the article on experimental physics d'Alembert wrote for the great encyclopedia that he edited with Diderot.[54]

The article begins with an intriguing addition to the standard distinction between observation and experiment. Experiment, d'Alembert wrote, is the more intrusive mode of investigation and enables one to study "the physics of the hidden (*occult*)." He was not, he hastened to add, proposing a revival of natural magic and explanations that refer to hidden spiritual forces, he was merely pointing out that the use of experiments carries one beyond the realm of passive observation by making available "knowledge of hidden facts which one has seen."

D'Alembert constructed a history of the experimental method that weaves together three themes: an interpretation of Newton's contributions to physics, a generational account of revolutionary change in science, and an enlightened critique of science teaching in French universities. His comments on how new ideas come to be accepted in science deserve special notice, for Lavoisier later described the reception accorded to his novel chemical ideas in exactly these terms. Many twentieth-century students of scientific change also hold similar views on the processes by which new theories are assimilated into science.[55]

> The generation that opposed [the early experimentalists] died off in the academies and in the universities, which today take their cues from the academies. A new generation arose, for when the foundations of a revolution have been laid it is almost always the next generation that completes it. It is seldom completed earlier because obstacles decay rather than perish; it is seldom completed later because once the walls have been breached, the human mind usually rushes ahead until it encounters a new obstacle that forces it to stay in one place for a long time.

The revival of experimental physics culminated with Nollet's appointment to a new chair in the University of Paris, an event which occurred shortly before d'Alembert wrote his article and at the height of the Buffonian attack on Nollet's electrical theories.

> These are the circumstances in which the king decided to establish a chair of experimental physics in the University of Paris. The current state of physics among us, the interest that those who know nothing of the subject show for it, and the example of foreigners who have long enjoyed the advantages of such an arrangement, all evidently required that we seek something comparable. Conditions had never been better for introducing a taste for well-grounded physics into as useful and distinguished an institution as the University of Paris, and it has met with considerable success for several years. The fame of the academician who occupies this chair is equal to the distinction with which he fills it.

Lavoisier could not have asked for a more authoritative or forthright endorsement of the scientific method he adopted as his own.

D'Alembert then went on to discuss the subjects to which the methods of experimental physics should be applied, and here too he provided an explicit warrant for the research program that Lavoisier was to follow.

> The true and perhaps unique goal of experimental physics [is to study] the phenomena, infinite in number, whose causes cannot be known by reason, whose connections cannot be immediately perceived, or whose connections can be seen at best imperfectly and infrequently and only after examining the phenomena from all sides. Such are, for example, chemical phenomena and those of electricity, the magnet, and a great many more. These are the facts which the physicist must try hardest to understand.

D'Alembert espoused and helped legitimate the conjoining of experimental physics and chemistry. He also commended the theoretical restraint advocated by Nollet and practiced by Lavoisier.

> The goal [of the experimental physicist] is to arrange [the phenomena he studies] in the best possible order, to explain some phenomena in terms of others in so far as is possible, and to construct a chain, so to speak, in which there are as few missing links as possible. That is enough, for nature is orderly. One must guard against the inclination to explain things that cannot be examined and against the frenzy for comprehensive explanation that Descartes introduced into physics. His disciples have become accustomed to vague principles and reasons that are equally capable of supporting the case for or against the proposed explanation.

To avoid the pitfall of explaining with vague principles, d'Alembert counseled, cultivate a reliance on the concreteness of facts.

> In physics courses explanations should be like reflections in a history – brief, profound, precise, and either introduced by facts or presented in a manner that surrounds them with facts.

In the preface to his *Metaphysical Foundations of Modern Science*, the German philosopher Immanuel Kant insisted that mathematics is the touchstone of genuine science: "I maintain that in every special natural theory one can encounter genuine science only to the extent that one encounters mathematics in it." He argued further that, since even the most general statements of chemistry rest on nothing more than empirical foundations rather than on *a priori* axioms, chemistry can never become a "genuine science."[56] Lavoisier, as a student of Lacaille and Nollet, thought otherwise. While taught to esteem the rigor of mathematics, he never made his understanding of chemistry hostage to mathematization. He treated the theory of attraction in the same way. Newton's achievement was not to be denied, but as both Nollet and d'Alembert cautioned, his theory's success in astronomy provides no warrant for assuming it can be readily or successfully applied to other phenomena.[57] Lavoisier therefore felt free to use the language of attraction when describing chemical phenomena without thereby committing himself to Newton's or Buffon's theories. He did not consider his chemical theories inadequate just because they could not be intimately integrated into Newtonian natural philosophy.

To appreciate the role mathematics played in Lavoisier's chemistry, one must distinguish between quantification, which he pursued with singular vigor, and the use of mathematical expressions as the language of theory, which he largely eschewed.[58] Lavoisier did not view this distinction as evidence of a radical disjunction in the forms of knowledge, for, according to Lacaille, the structure of mathematics itself emerges from the quantitative investigation of natural phenomena. Lavoisier thus thought of mathematics as a highly formalized way of stating empirically based knowledge. It was in this sense that mathematics provided an ideal to which all the sciences aspired.

According to Lacaille, mathematics, rather than being a self-contained system of thought one applies to nature, is a way of reasoning that emerges from attempts to quantify different realms of science.[59] Mathematics is the logic of quantification and every science quantifies what it studies in a way appropriate to its particular subject-matter.

> Everything that can be increased or diminished is called a quantity. The numbers appropriate to, for example, extension, movement, light, etc. have different quantities. Mathematics encompasses all those sciences that examine the properties and relations of these diverse quantities. Each of these sciences has its own distinctive name, which accords with the nature of the things its studies. Arithmetic is the science of numbers; geometry is the science that measures and compares the three dimensions of extension – length, width and depth. The science of movement

and equilibrium is called mechanics, the science of light is called optics, and so it is with others.[60]

Lacaille taught Lavoisier that mathematics in its formal aspects is a rigorous science and the model for all deductive reasoning and demonstrative proof. But when considered together with the empirical investigations from which it emerges, mathematics, like the other natural sciences, can be seen to be grounded in the study of physical nature. The physical scientist should pursue quantification and trust that, as well-established quantitative knowledge increases, the theories it gives rise to will become more rigorous.[61]

Lavoisier was perfectly forthright about his views on mathematics and its relation to the study of nature, most notably in the preface to his *Elementary Treatise on Chemistry* (*Traité élémentaire de chimie*). As he wrote,

> The only way to prevent errors [based on suppositions] is to suppress reason, or at least simplify it to the greatest extent possible, for it comes entirely from us and if relied on can mislead us. Reason must continually be subjected to experimental proof. We must preserve only those facts that are given by nature, which cannot deceive us. Truth must only be sought in the natural connection (*enchaînement*) between experiments and observations, in the same way that mathematicians arrive at the solutions of a problem by a simple arrangement of the givens. By reducing reason to the simplest possible operations and restricting judgment as much as possible, they avoid losing sight of the evidence that guides them.[62]

Elsewhere in the preface Lavoisier cleverly freed himself from the difficulties associated with mathematics and Newtonianism by elevating them to the realm of the transcendent. He began by explaining why he would have little to say about affinities:

> The rigorous law from which I have never deviated, of forming no conclusions which are not based on experiment and of never interpolating in the absence of facts, has prevented me from addressing in this work the branch of chemistry that is most likely to become an exact science, namely that which deals with chemical affinities and elective attractions.[63]

He was reluctant to dwell on affinities, Lavoisier declared, solely because the subject had not been sufficiently developed: "The principal facts are not known and those that we have are either too imprecise or too uncertain to provide a foundation for such an important branch of chemistry." Then, having protested almost too much, he cut free from the threatening entanglements of the higher philosophy.

> The science of affinities is to ordinary chemistry as transcendental geometry is to elementary geometry. I thought it unnecessary to add this complication to the simple and easy elements that I hope will serve as an introduction to many readers.[64]

Let us return briefly to Buffon to see how he responded to Lavoisier's new chemistry. In 1788 Thomas Jefferson, who was then living in Paris, summarized his views in a newsy letter to James Madison.[65] Buffon had recently died, but the opinions he expressed in his final years continued to command respect. "Speaking one day with Monsieur de Buffon on the present ardor of chemical enquiry," Jefferson wrote, "he affected to consider chemistry but as cookery and to place the toils of the laboratory on a footing with those of the kitchen." Jefferson objected, he told Madison, and insisted that chemistry is far too useful to be dismissed in this manner. Yet it is true, he added, that

> it's principles are contested. Experiments seem contradictory: their subjects are so minute as to escape our sense and their result too fallacious to satisfy the mind. It is probably too soon to propose the establishment of system. The attempt therefore of Lavoisier to reform the Chemical nomenclature is premature.

Buffon, never one to change his mind in the face of criticism, remained true to his beliefs to the end. By doing so he blinded his friend Jefferson to the virtues of a scientific revolution he should have welcomed. What we have here, I suggest, is not merely a difference of opinion about alternative theories or paradigms, but rather a more fundamental confrontation between distinct scientific styles. If Lavoisier's scientific style was in some ways bourgeois republican, then Buffon's was aristocratic. Both men were members of the Academy of Sciences and had to abide by its codes of conduct. But Buffon, unlike

Lavoisier, turned away from the Academy and directed his energies to building up the King's Garden, which he ruled with the autocratic hauteur of a grandee. The King's Garden was Buffon's court. While dividing his time between Paris and his country estate and while producing his great compendium of natural history, Buffon played the role of a prince in the realm of eighteenth-century French science. The scientific court he created and patronized was composed of admirers, informants, assistants, and subordinates, and he commanded them as their lord rather than seeking to persuade them as a fellow citizen. Its members contributed to his vast enterprise but did not shape it. Buffon proclaimed interpretations and explanations that he was not inclined to revise. He was an exemplary aristocrat of science in an aristocratic age; how fortunate for him that he died a natural death just before the world in which he had flourished was brought to ruin.

4

The Chemistry of Salts

In the eighteenth century the Academy of Sciences periodically provided evidence of its privileged status by engaging in elaborate public displays. These ceremonies were something of an innovation, for in the seventeenth century all Academy meetings were held behind closed doors. It had been realized, however, that such inwardness did little to promote public recognition of the Academy's broad authority within its domain of expertise. The regulations of 1699, which the Parlement of Paris registered in 1713, were therefore sanctioning evolving practice when they required the scheduling of certain special meetings. These public sessions were held following the semi-annual vacations, one in the spring near Easter and the other in the fall near St Martin's Day. On these occasions selected members of the Academy described their research, new members were presented, and those who had passed on were eulogized by the secretary. The gossipy Parisian news journals provided full reports of these notable social events.[1]

Late in 1772 Lavoisier asked that he be allowed to describe his most recent findings at the next public meeting of the Academy. He believed that certain experiments on the nature of combustion and calcination had led him to a discovery of considerable importance. Surviving manuscripts indicate that he initially hoped to present these experiments at the public meeting in the fall of 1772, but he was unable to obtain a place on the program. He therefore immediately outlined his discovery in a note that he had the Academy's secretary initial and date. Having thereby protected his interest, Lavoisier returned to his experiments. During the next few months he extended and refined his new ideas while preparing for the next public meeting.

The paper he presented at the Easter meeting in 1773 was titled "On a new theory of the calcination of metals." It is unlikely that many of those who heard his talk gave it their undivided attention, for the experiments he described were not in themselves dramatic and their theoretical significance was far from obvious. But with the advantage of hindsight we can see that Lavoisier was then providing a first public account of his striking new insights into the chemistry of fixed air. And it was his pursuit of this line of investigation that led directly to the momentous theoretical innovations for which he is still remembered.[2]

In his talk at the 1773 Easter meeting Lavoisier pointed out that iron and copper, when exposed to air, are slowly reduced to the powdery substances called rust and verdigris. It had not been widely noticed, however, that as the metals pass from the metallic state to the powdery state (i.e. are calcined and become calxes), their weight is sensibly increased. This interesting fact, he reported, had not yet been adequately explained. Lavoisier's recent experiments indicated that "there is an absorption of air during calcination and a disengagement of air during reduction." He therefore concluded that "(1) a metallic powder is formed by the combination of a metal with a certain quantity of fixed air, and (2) metallic reduction consists essentially of the disengagement of the air with which the metal has in some manner been saturated." He then added that, since it is the abundant fixed air of the atmosphere that causes weight gain in calcination, he will ignore the question of whether Stahl's phlogiston also enters into the process.[3]

Lavoisier's recognition that the weight increase during calcination is caused by the fixation of air marks a turning point in the development of his chemical research program. Prior to that he had addressed a variety of problems concerning the nature of the elements and, more particularly, the fact that air, fire, and water can exist in two different states, the fixed and the free. His discovery of the role of air in calcination focused Lavoisier's research and set him on the trail that led to the oxygen theory of calcination, combustion, and acidification. Of course he could not simply deduce his mature theories from this experimentally based insight; they had to be hammered out through years of careful experimental investigation and arduous conceptual revision. In retrospect, however, we can see that by the end of 1772 Lavoisier was pursuing the research program that would bring him great personal glory and transform the science of chemistry.

How did Lavoisier advance from a general desire to apply the methods of experimental physics to chemistry to a highly focused research program that made the properties of air fundamental to theoretical chemistry? This is the central question addressed in this chapter. The answer to it lies in the chemical concepts and analogies Lavoisier employed as he struggled to make sense of the subjects he investigated. Our story therefore begins with the organizing doctrines Lavoisier acquired while studying with his earliest chemistry instructors and ends with his presentation of the Easter memoir of 1773. Above all else it is a story of physical elements, how they were defined and how they were thought to give rise to the observable properties of substances.

Guillaume François Rouelle was one of the most popular public lecturers in Paris. Born in 1703 and educated in Caen, he moved to Paris in the 1730s, where he served as an apprentice to a German pharmacist. By 1740 Rouelle was giving lectures on chemistry and pharmacy and in 1742 Buffon appointed him as demonstrator in chemistry at the King's Garden. He was granted unusually generous terms, considering that he was not the professor of chemistry, and was allowed to give independent lecture courses, which like all courses taught at the King's Garden were open to the public.

Rouelle was also a serious and productive investigator. In 1744 he published the first of a series of important papers on the classification of salts and was admitted to the Academy of Sciences. Two years later he began giving private courses in his new laboratory on the rue Jacob. By 1750 he had succeeded in establishing himself as the foremost chemical teacher in Paris. The experiments with which he illustrated his lectures were not entirely novel, but he carried them off with an enthusiasm and flamboyance that inspired his auditors. To provide some theoretical coherence to his course he developed a distinctive set of doctrines that synthesized G. F. Stahl's phlogistic chemistry and Stephen Hales's studies on the fixation and release of air.

Rouelle was truly the instructor of the generation that succeeded him. Prominent chemists who studied with him included, in addition to Lavoisier, Gabriel François Venel, the author of over 700 articles on chemical topics in the *Encyclopédie*, Macquer, Bayen, d'Arcet, and Desmarest. Prominent *philosophes* who are not remembered as chemists also attended Rouelle's courses; among them were Denis Diderot, who sat through his course three times in the 1750s, Baron d'Holbach, Jean Jacques Rousseau, the administrator A. R. J. Turgot, and the revolutionary leader Malesherbes. Rouelle continued to lecture through the 1760s, although his health began to fail in the latter part of the decade and he died in 1770. Lavoisier was fortunate to catch him in his prime.[4]

By the time Lavoisier began studying with Rouelle chemistry had become a popular, well-established, and highly respected branch of science in France. This had not been the case at the beginning of the eighteenth century, when chemists worried that their subject might be swallowed whole by the comprehensive systems of natural philosophy championed by Cartesians and Newtonians. This fear had prompted many prominent chemists to insist that their science studied a distinctive set of properties of matter and that their theories could not be reduced to mathematics or mechanics. Stahl's system of chemistry was popular in large part because it argued strongly for the disciplinary autonomy of the science. Shortly after the middle of the century G. F. Venel still found it necessary to insist that chemists study distinctive properties of nature and hence must create theories of their own.[5]

By the 1760s the notion that chemistry might be reduced to a branch of physics had largely been dispelled by Rouelle's popular success and by a growing disenchantment with comprehensive theories of nature. The discipline Lavoisier encountered in Rouelle's lectures was therefore autonomous and self-confident, but it was still quite an incomplete science.[6] As we have seen, Lavoisier had not been impressed by the chemical theories he was introduced to as a student, but he seems not to have questioned the importance or independence of chemistry itself.

Its place in the house of science was secure; his task was to improve the theories with which chemists explained certain phenomena that lay in the borderland between chemistry and physics.

Rouelle was evidently comfortable with the recognition accorded to chemistry and, while he introduced his auditors to the traditional distinction between chemistry and physics, he did not consider the two sciences competitors. In an attempt to bring them into a closer alliance he organized his lectures around an accommodating theory of elements. Chemists had long debated over the nature and number of the elements. Rouelle thought there were probably five, but he found the ancient theory of four elements served his purposes best. This theory had the advantage of providing a common set of elements for both physics and chemistry. The elements or substantial principles, as he also called them, were phlogiston or fire, air, water, and earth. His characterizations of these elements reflected his epistemological caution. While Rouelle did indulge in some transduction, that is to say he occasionally spoke about the essential properties of unobservable entities, he understood that all theories of elements are fundamentally heuristic. As he told his students, "all one can say with some assurance is that there is a very small number of them and that the different ways in which they can combine are capable of producing all the substances found in nature."[7]

Chemists could not examine the elements themselves and therefore had to draw inferences from the properties of higher-level compounds that the Stahlians called "mixts." Chemists acquired knowledge of the elements by analyzing mixts into their constituent parts and by synthesizing substances and forming new mixts. By performing these operations on different substances in a variety of circumstances, chemists could identify the enduring and hence causal properties of the different elements and their compounds. As Rouelle acknowledged, however, this way of proceeding can at best provide probable rather than certain knowledge.

> If on occasion we succeed in analyzing substances, their parts immediately attach themselves to other substances and form new mixts or new compounds. Chemistry draws most of its insights from these new combinations. We determine the properties of the principles of a mixt by following them as they move from one combination to another. By allow-

ing the principle to pass successively through several combinations, we can see what is indestructible, immutable, and indivisible. But these operations can only provide evidence that it is difficult to deny.[8]

Rouelle followed traditional practice in calling each of the elements a natural instrument, his point being that they are all active agents of chemical change. The instrumental properties of fire and water, long used by chemists to analyze substances in the dry and the wet ways, were especially important. By the middle of the century, however, this use of the term instrument was becoming a bit archaic.[9] More significant for the future of chemistry was Rouelle's use of instrumental apparatus in the laboratory. Experiments played a central role in Rouelle's chemical investigations and in his presentation of his findings, and Lavoisier no doubt admired this aspect of his practice.

Rouelle's discussion of elements throws considerable light on some of the most distinctive features of Lavoisier's mature chemistry. Like Rouelle, Lavoisier treated the identification and characterization of the chemical elements as a matter of investigative tactics rather than philosophical ontology. He posited the existence of simple substances because he had to have some "things" to reason with, not because he believed he could gain absolutely secure knowledge about the ultimate units of matter. Earth, air, fire, and water provided reasonable starting points for physicists and chemists, yet Lavoisier was prepared to consider the possibility that water, for instance, can be turned into earth or air. The kind of knowledge science builds on is experimental and quantitative. The simplest substances that can be known are those that resist all further attempts at analysis. In principle, therefore, Lavoisier was prepared from the outset to accept the plurality of airs and the analysis of water into oxygen and hydrogen. He had learned from Rouelle to treat theories of elements heuristically. When formulating his own theories, he was fully aware that the "things" he called simple substances were historically and instrumentally contingent.

Lavoisier also followed Rouelle in thinking of chemical elements as the bearers of chemical properties. Here too he was treating the elements heuristically. He knew he could not get behind the available evidence to explain how the imperceptible elements cause perceptible properties. Cartesian corpuscular motions, Newtonian attractive forces between atoms, and the molecular shapes posited by Lemery and Buffon were all equally inaccessible. Rouelle simply assumed that properties that regularly appear when certain elements are present are caused by those elements. Lavoisier saw no reason not to follow the same strategy. Consider, for instance, Lavoisier's theory of acidity. Sulfuric and phosphoric acid are made by burning sulfur and phosphorus in the atmosphere and then dissolving the products of combustion in water. Since combustion involves the fixation of oxygen, Lavoisier made the reasonable inference that oxygen is the cause of acidity. This explanatory claim appeared to be justified by the available evidence but was nonetheless subject to future correction. Of course Lavoisier, like anyone having a vested interest in an idea, was not inclined to abandon his oxygen theory of acidity when it was first challenged. In principle, however, even this fundamental tenet of his new chemistry was corrigible.

All this agonizing over the possibility of acquiring reliable knowledge about elements and causes would have been fruitless had Lavoisier not had reason to believe he could get beyond mere hypothesizing. Rouelle had encouraged him to believe that the systematic study of chemical changes would prove rewarding, but what really got him fired up were changes of state. How could certain substances, and especially the elements water, fire, and air, exist as solids, liquids, and vapors? The question had been asked before, but recent studies of the fixation and release of air had given it a new urgency. Here was a subject of interest to both physics and chemistry; it was also one whose investigation required the use of precise instruments and the collection of accurate data. Lavoisier could hardly wait to get going.

Eighteenth-century physicists and chemists had already devoted considerable attention to changes of state.[10] These changes appeared especially dramatic because several of the elements, when fixed, lost what were thought to be their essential qualities: fixed air, such as the air in limestone, is not elastic; fixed fire, such as latent heat, is not hot; and fixed water, such as the water in crystals, is not fluid. The dramatic differences between the properties exhibited in the fixed and the free states made it difficult to defend physical or chemical theories founded on the classical "essential properties" of these elements. Changes of state thus provided investigators with interesting and clearly demarcated changes in the properties of elements and compounds that could be examined and quantified in the laboratory. Lavoisier recognized all this and eagerly pursued the opportunities available to him. He returned again and again to the study of changes of state, his attention always being focused on experimental quantification. Although theoretical command of these topics did not come easily, he persevered and eventually triumphed.

The path Lavoisier followed was surprisingly well prepared and ran along ground of interest to both chemists and physicists. In the memoir on gypsum he read to the Academy in February 1765, Lavoisier mentioned and commended Nollet's paper on the boiling of liquids, a memoir the Academy published in 1748.[11] Nollet had been concerned with the standard problem of explaining the formation of vapors. Is air the only permanently elastic fluid? Do liquids dissolve in air or do they form vapors by combining with heat? Recent experiments on the boiling of liquids in evacuated chambers led Nollet to favor the heat theory. Indeed, he found that the vapor formed in boiling "imitates the air so well in certain regards, that I have been tempted to take it for the same thing."[12]

Nollet's readiness to practice what he preached could not have been more forcefully demonstrated. If his experiments indicated that vapors are formed by heat and if those vapors behave just like air, then he was prepared to say that air too is a vapor formed by heat. In citing and commending this memoir Lavoisier indicated that when discussing the elements he too would go where his experiments led him. In doing so he was simply conforming to the expectations of the academicians he was trying to impress. As Nollet wrote in 1769, when replying to a correspondent who had sent him a paper full of speculations based on the properties of the ultimate particles of matter, "I ought to let you know that the Academy is becoming increasingly opposed to this type of philosophizing."[13]

Lavoisier was also impressed by the way A. R. J. Turgot dealt with changes of state in an article he wrote for the *Encyclopédie*.[14] Turgot suggested that air, rather than being called essentially elastic, should be described as "expansible." According to this reconceptualization, air is like alcohol and the other liquids that combine with heat when they are "vaporized," another term invented by Turgot. Turgot introduced these terms to distinguish between expansible substances that are vaporized by combining with heat and those, such as the water vapor normally found in the atmosphere, that evaporate by slowly dissolving in air. This way of viewing the problem was not unreasonable, as Turgot pointed out, for water, depending on its temperature, can exist as a solid, liquid, or vapor. By 1772 Lavoisier was familiar with this article. He made considerable use of Turgot's new terms and they no doubt encouraged him to concentrate even more intensely on the theoretical problems that Turgot had addressed.

When seeking to make sense of his earliest chemical investigations, however, Lavoisier was understandably conservative and relied primarily on Rouelle's theory of salts. The chemical investigation of salts had a complicated history that reached far back in time; it was also a topic of particular interest to the experimental chemists in the Academy. What captured Lavoisier's attention was the central role that changes of state played in Rouelle's theory of salts.[15] Rouelle was in fact more concerned with classifying salts than explaining their properties. He defined a neutral salt as "a salt formed by the union of an acid with any substance which serves as a base and gives it a concrete or solid form." This definition addressess two different kinds of changes. Chemically, it characterizes the formation of salts as an acid-alkali neutralization, while physically it refers to a change of state from liquid to solid. This was a linking of phenomena that Lavoisier was determined to explore.

Many chemists had attempted to find a quantitative measure for the relative strengths of acids and alkalis, but with little success. In 1700 the academician William Homberg developed a technique for using equivalent weights to measure the relative combining powers of different acids. He began with a fixed weight of a certain alkali salt and measured the weight of a particular acid required to neutralize it. The success of the technique depended on the correctness of the assumption that the acid and alkali were the only chemically active agents in the reaction. As Lavoisier no doubt realized, however, the effervescence given off during neutralization clearly indicated that Homberg had not succeeded in capturing and weighing all the substances involved.[16]

Rouelle approached the study of salts by examining the crystals they form. He used crystal shape to distinguish among six "sections" or types of neutral salts. Within each section he identified the genera of each salt by its acid and its species by its base. Further study enabled him to distinguish between acid salts, which have an excess of acid, and neutral salts, which have just enough acid to achieve saturation. He also distinguished between the water of crystallization, which enters into the formation of the crystal, and the water in excess of that needed for crystallization, which can be evaporated off without destroying the crystal. As we shall see below, Lavoisier was at the outset more impressed by the fixation and release of water in the formation of salt crystals than by the role of air in chemical change. He believed that by focusing on this set of physical changes he would be able to extend Rouelle's theory of salts into other areas of chemistry. This would enable him to infuse the study of chemical elements and their properties with the rigor of quantitative experiment.

The memoir on gypsum Lavoisier read to the Academy in February 1765 provided ample evidence of the candidate's extraordinary self-confidence.[17] Although only twenty-two years old, he boldly announced that this, his first scientific paper, marked the beginning of a major research program devoted to the investigation of earths, minerals, and crystals. But rather than employing the traditional technique of analysis by fire, he would copy nature by exploiting the properties of water, her principal agent and a solvent capable of rendering almost everything liquid. Clear and compelling experiments would provide an indisputable foundation for the truths to be demonstrated.

Lavoisier began his analysis of gypsum by heating a sample of the mineral in an iron dish. "During this operation one notices a vapor, a light smoke, is driven off."[18] Lavoisier collected this vapor and found it to be pure water. It is the water of crystallization, he reported, and constitutes about one-quarter of the weight of the gypsum. When this water is remixed with the roasted gypsum, crystals form and the gypsum again becomes a hard solid mass. The water that combines with the salt in gypsum is therefore the cause of its solidity.[19] The truth of these claims, he insisted, are demonstrated in his memoir.

The main purpose of Lavoisier's first memoir was to demonstrate that gypsum (hydrated calcium sulphate) is a true neutral salt that becomes a solid by fixing water and forming crystals.[20] His second memoir on gypsum, which he read to the Academy just over a year later, was, as he termed it, an analytic history of the various kinds of gypsum found near Paris.[21] He began this second memoir with a fulsome acknowledgement of his debt to Rouelle.

> There is a certain sequence in human knowledge that cannot be broken and that is crucial to the success of our discoveries. So it is that Rouelle's renowned doctrine concerning the different quantities of acid that can combine in the formation of the same salt necessarily had to precede an account of the various forms of gypsum. This most fruitful discovery, the greatest advance in theoretical chemistry since Stahl, provides the basis for all the causal explanations presented in this memoir. Indeed, I have no doubt that it will serve in the future as the basis for a great many more and that posterity will find that it renders intelligible the most impenetrable mysteries of nature.[22]

Lavoisier believed Rouelle's theory of salts would explain why different minerals combine with water to form a variety of crystals.[23] He decided not to investigate the formation of vapors under the action of heat because Nollet had already given an adequate account of boiling in his 1748 memoir.[24] At this stage in his investigations, therefore, Lavoisier decided to focus on the relationship between the chemical composition of salts and the fixation of water in solids.

During the year that elapsed between the presentations of his two memoirs on gypsum, Lavoisier devoted a great deal of time to his memoir on street lighting. In 1765, while puzzling over the many practical problems involved in designing, constructing, and operating different types of lamps, he examined the properties of the available fuels.[25] Oil pressed from seeds, especially olive oil, appeared to be the fuel of choice. The main alternative, neat's-foot oil, was in short supply and congealed more quickly in cold weather.

Lavoisier found the problem of congelation intriguing and performed a series of experiments on the olive oil used as lamp fuel. One conclusion he reached is a vague formulation of the difference between the intensity and the quantity of heat: "It is not the cold's intensity that congeals oils, rather it is extended exposure to cold of that intensity."[26] But why do oils congeal at different temperatures? Chemists had determined that oils pressed from seeds contain certain resins extracted from the seed covers. The analogy of salt crystallization suggested that these resins might retard congelation.

84

It is known that when oils pressed from seeds reach a certain temperature they dissolve resins. In this respect they act just like water on salts:
solution and crystallization are common properties of these two liquids.
If salts dissolved in water delay freezing, why not assume the same thing
happens when resins are dissolved in oils?[27]

Lavoisier realized that such an explanation remained conjectural
until confirmed by experiments. The analogy was suggestive, however,
and his use of it reveals once again that from the very beginning of his
career he was preoccupied with the chemical theory of salts, with
changes in state caused by the fixation and release of air, fire, and
water, and with the connections between heat and changes of state.

As early as the end of spring in 1766, a few months after presenting
his second memoir on gypsum, Lavoisier was beginning to appreciate
just how problematic his research program was. Although he continued to treat Rouelle's theory of salts and Nollet's theory of vapors as
well-established starting points, he realized that the identity of the
elements themselves remained an open question. He had hoped that by
adhering to the method of experimental physics he could avoid being
drawn into speculative debates over the nature of phlogiston or the
number of earths. Like most French chemists of his generation, he
simply assumed that for heuristic purposes he could treat water, air,
fire, and the earths as distinct and enduring forms of matter. But two
memoirs he read sometime before the end of May 1766 made him
wonder if he was building on sand.

In 1750 the Berlin philosopher J.T. Eller published two papers, one
on water as a solvent and the other on the solution of salts in water.[28] It
was probably after reading these memoirs that Lavoisier came across
and read Eller's earlier memoir on the elements. Eller hypothesized
that there are only two elemental principles, water and fire. Air, he
asserted, is a combination of the two. Fire, according to Eller, is the
active agent in all solvents. The sun is the ultimate source of all the fire
fixed in solvents, including the principle of acidity that is concentrated
in the most powerful of all solvents, the mineral acids.

In May 1766 Lavoisier wrote two brief notes summarizing his reactions to Eller's hypotheses.[29] The first of these notes has three headings:
"physical chemistry," "on the elements," and "on fire, water and air";
the second is headed "chemistry" and "on the matter of fire and the
elements in general." When interpreting these manuscripts we must

remember that they were notes Lavoisier made for his own purposes. They allow us to watch him closely as he plays with ideas, but they do not represent his final conclusions on the questions addressed.

In the first of these notes Lavoisier puzzled over Eller's theory of air, which seemed identical to Nollet's, and wondered how the solar fire that combines with water to form vapors enters into the compound.

> Air is not an element that exists by itself, it is compound. It is water turned into vapor. To put the point more clearly, it is the result of the combination of water with the matter of fire. It remains to be seen whether it is the pure solar fire of the atmosphere that unites with water or whether that same matter enters into combination in the form of a universal acid (*acidum pinque*).

The second note begins with a Cartesian speculation that no doubt reflects the philosophical predilections Lavoisier acquired while studying with Nollet.

> I accept that there is a solar atmosphere that surrounds the entire planetary system. Just as the air is the medium that transmits sound, the solar atmosphere is the medium that transmits light. There is no reason to insist that light is an emanation of the solar substance itself. The famous argument of the Newtonians, that light is successive transmission, proves nothing.

As Lavoisier mulled over these ideas the concept of a solar atmosphere threatened to obliterate the very distinctions that made the theory of elements useful.

> All the elements, air, water, and earth, are contained in the primitive fluid of this solar atmosphere. They dissolve one another reciprocally, so that the air contains both the igneous fluid and water in solution, just as this last element holds these two elements in solution to the point of saturation.

Lavoisier plunged on, attempting to clarify these slippery concepts with an even more speculative Cartesian discussion of the sizes of air, water, and fire particles. But then, towards the end of this brief note, he recovered his poise and asked, "isn't air itself a fluid in expansion?" And just before concluding this series of speculations he was once again restricting himself to observable properties and, more particularly, those exhibited in changes of state.

> Everything is composed of the elements, but they don't always combine together in the same way. Air and fire, when in the fluid state, show considerable similarity in this regard. They both lose some of their essential properties when they form compounds. It is known, for example, that when air ceases to be elastic it occupies infinitely less space than when free.

Shortly after jotting down these reflections Lavoisier learned that he had not been elected to the Academy. Perhaps in reaction to this disappointment, he immediately threw himself full-time into his and Guettard's geological mapping project. He traveled with Guettard for months on end and evidently was happy to forget, while engaged in field work, the irresolute debate over the number and nature of the elements. August of 1767 found them near the Swiss border in Alsace while traveling from Basle to Strasbourg. Lavoisier wrote enthusiastically to his father of the local dairymen's hardihood and of the pleasure he took in climbing in the Vosges mountains.[30] His scientific work, like that of his mentors Lacaille and Nollet, was now organized around the daily use of instruments. He fretted over the fragile barometers with which he measured elevations, measurements he then entered on the margins of their maps. He also recorded thermometric readings and regretted that he had to forgo other opportunities to make instrumental observations.

> A great many interesting physics experiments could be made in a region such as this. Had I thought of them before leaving Paris, I could now take into consideration additional phenomena that I must neglect because I do not have the necessary instruments. Electricity is one such subject and I have no doubt that if one were to perform experiments, one could discover important truths about how storms arise.[31]

Lavoisier did not neglect chemistry while engaged in field work. He tirelessly analyzed the different waters they came upon in their travels and looked forward to reporting on his efforts upon their return home. As he told his father, "when I get back to Paris I will have made complete analyses of samples of all the natural and mineral waters of this region."[32] Reinvigorated by his time away from the city and recon-firmed in his belief that the investigation of physical nature was best pursued through the use of instruments, he returned to the Academy convinced that his original research program, however much it might have to be modified, was fundamentally sound.

The first memoir Lavoisier presented to the Academy upon his return gave ample evidence of his unwavering commitment to the method of experiment. In January 1768 be began a series of investiga-tions using hydrometers to measure the density of water at different temperatures.[33] Once again Lavoisier was attempting to carry further a mode of analysis already being used by his colleagues in the Academy. He employed both variable immersion hydrometers, of the sort that are used today to test the strength of battery acid and engine coolant in automobiles, and constant immersion instruments. The instrument's cylindrically shaped body was designed to float upright in the liquid whose density was being measured. The constant immersion variety was calibrated by adding weight to the bottom of the instrument until it just barely floated in distilled water, at which point a mark was made on the neck. To measure the relative density of another liquid, one floated the hydrometer in the liquid and added weights to the small cup on top of the neck until the calibration line was at the surface. The amount of weight added then enabled one to calculate the relative density of the liquid.[34]

Lavoisier described his hydrometers and reported on his findings in the paper he read to the Academy in March 1768.[35] Determining the specific densities of liquids was, he declared, a problem of central importance to physics.[36] But as he indicated towards the end of his paper, the techniques he employed to solve this problem would also be of interest to chemists.

> Chemists have plenty of ways of precisely determining the quantities of solid or concreted matter they employ in their experiments. The balance provides reliable tests that cannot be mistaken. But there are problems with certain salts that cannot be reduced to a solid form, as is the case with most acids, especially the mineral acids. The balance provides us with the sum of the weights of the water and the parts of the salts that form these fluid acids, but it does not tell us in what proportion they are mixed. It is on this point that the hydrometer can be wonderfully helpful.[37]

Lavoisier proposed that the data collected through the use of hydrometers be organized into tables. Evidently he thought that the strength of the various acids was directly proportional to the relative densities of their solutions. It also appears that he thought the property of acidity is caused by a single element and that strong acids contain more of this element than weak acids, although he did not state this hypothesis explicitly. Clearly he once again hoped that by finding an experimental means to measure a physical property associated with an important chemical property, he would be able to make an advance toward the goal of transforming chemistry into a precise science.

After claiming this new beachhead for instrumental chemistry, Lavoisier listed some of the many problems he believed could be resolved by the skillful use of hydrometers.

> It is primarily the art of combination that can be illuminated by knowledge of the specific weight of fluids. This part of chemistry is much less advanced than is thought; we hardly know the first elements. Every day we combine acids and alkalis, but how do these two substances unite? Do the constituent molecules of the acid enter the pores of the alkali, as Lemery thought, or do the acid and alkali have different facets that somehow engage one another or simply unite on contact, like the hemispheres of Magdebourg? How are acids and alkalis held separately in water? How are they held after they combine? Does the salt that is formed simply occupy the pores in water? Is there simply a division into particles, or is there a real combination, whether between individual particles or between one particle and many others? Finally, where does the air that escapes with such liveliness at the moment of reaction come from, air which, when in its natural state of elasticity, all at once occupies a volume enormously greater than that of the two fluids from which it emerges? Does this air exist from the outset in the two compounds? Was it in some way fixed, as Mr Hales and most physicists since have supposed, or is it so to speak a factitious air that is, as Mr Eller thought, a product of the combination? Chemistry, when asked to address these possibilities, answers with empty terms, such as similarities, analogies, and frictions . . . which clarify nothing and encourage the mind to be satisfied with words alone. If it is possible for the human mind to penetrate these mysteries, it can hope to do so through research on the specific weights of fluids.[38]

This prescription for the advancement of chemistry is surprisingly frank. Lavoisier, who was still a candidate for election to the chemistry section, openly declared that, if chemistry was to advance, it must adopt the instrumental and quantifying program of experimental physics. Use of the balance to determine weight changes was already a central part of chemistry; what he had developed is a technique for quantifying changes that cannot be measured directly by the balance. The young man certainly knew his own mind and did not lack courage.

Lavoisier discussed chemical modes of analysis with somewhat greater circumspection in the paper he read to the Academy in May 1768.[39] This is the paper that contains the promised account of the water analyses he had performed while traveling with Guettard. Since water is nature's favorite agent, Lavoisier reasoned, the waters that occur in nature offer chemists a vast and fruitful field for analysis.[40] The trick is to know how to identify and quantify the substances dissolved in waters. Here again he attempted to use a physical means to solve a chemical problem.

The techniques favored by chemists are plagued by the difficulties involved in isolating elementary substances. The academician Le Roy read an interesting paper on this problem at the Easter meeting in 1767. According to Lavoisier, Le Roy demonstrated that "a considerable amount of earth can be carried over by the water in distillation, and this earth is separated" when salts are isolated by evaporation. If this happens to earths, which by definition are practically insoluble, it must occur to an even greater extent with salts.[41] Le Roy's more general conclusions must also have weighed heavily on Lavoisier's mind. He pointed out that while "nearly all physicists" accepted the ancient doctrine of four elements, the evidence for the conversion of water into earth posed a severe challenge to such beliefs.

> One can readily see that this assertion, if it were true, would upset all accepted ideas and would destroy, beyond recovery, all the certainty that one can expect of chemical analysis, because one could never be sure that the substances obtained by the decomposition of a mixture, instead of being the materials that were in it, were not the results of the process.[42]

Lavoisier went on to identify other difficulties that arise when one uses heat to separate water from the salts dissolved in it. He then brought his discussion of the analysis of waters to a close by praising and quoting the conclusion Macquer arrived at in his *Dictionary of Chemistry*. "And yet," Macquer wrote,

> despite all the careful effort that has been expended, it is evident that we are still far from attaining the certitude and knowledge that we would like to have concerning this important topic. This is not really surprising, however, for these are perhaps the most difficult analyses there are in chemistry.[43]

Challenged by this judgment, Lavoisier again claimed that his hydrometer provided a new and highly accurate way for determining the quantities of salt dissolved in liquids. And to demonstrate that his was more than a programmatic claim, he provided page after page of analytic data on the waters he and Guettard had examined while on tour.

Lavoisier's claims for his hydrometer were, of course, excessive, and it was not long before he realized he would have to find other ways to carry forward his instrumental investigations. He also realized that he had to resolve, at least to his own satisfaction, the question of whether the physical elements could be converted into one another by such processes as heating, fermentation, and the growth of plants. The question was of direct relevance to his research strategy, for if such operations do convert the elements into one another, then they could not be used to determine the quantities of salts, air, water, and earths contained in various compounds.

Late in the summer of 1768 Lavoisier attacked this problem with characteristic eagerness. As he wrote in his paper on the convertibility of water, "the question of the transmutability of the elements into one another, and especially that of changing water into earth, is exceedingly interesting for physics and has been discussed by a great number of famous authors." His goal, however, was to devote as little time as possible to speculation and move directly to experiment so that he could "speak of facts."[44]

According to Lavoisier, the belief that water can be transmuted into earth arose primarily from studies of vegetation. In the seventeenth century the chemist Van Helmont put two hundred pounds of dried soil in a pot, watered it, and planted in it a willow shoot weighing five pounds. For five years nothing but rainwater was added to the pot, yet the tree grew to a weight of 164 pounds while the soil lost only two ounces. It thus seemed reasonable to suppose that the great increase in the weight of the willow, much of which would have been earthy ash had it been burned, should be attributed to the vegetative transformation of water into earth. Other scientists, including J.T. Eller, had performed similar experiments and reached the same conclusion.

Lavoisier, eager as ever to clear away unexamined ideas, subjected this conclusion to stringent criticism. It was easy, of course, to point out certain inadequacies in his predecessors's experimental designs. More interesting, however, is his suggestion that earlier experimenters failed to notice a possible source of vegetative weight gain. The experiments of Hales, Guettard, Duhamel, and Bonnet had demonstrated that plants not only give off vapors, they also absorb vapors from the atmosphere. While making this point Lavoisier took the occasion to outline his vapor theory of air.

> One will say that if air is the source from which plants draw the various constituents (*principes*) that analysis reveals they contain, these same constituents should be present and identifiable in the atmosphere. To this I reply that, while we have no demonstrative experiments that this is so, one can hardly doubt that the lower part of the atmosphere, where vegetables grow, is highly complex. First, it is likely that the air that makes up the base [of the atmosphere] is not a simple substance or an element, as the first natural philosophers (*physiciens*) supposed. Second, this fluid is the solvent for water and all the volatile substances that exist in nature. I will soon be able to show, in a memoir I am preparing on the nature of air, that evaporation and dissolution in air are nearly the same thing. Finally, quite independently from the various volatile substances combined with and in some sense dissolved in air, salts and other highly fixed bodies are carried into it by water, although in small quantities.[45]

Having indicated he would soon provide a fully developed vapor theory of air, Lavoisier returned to the role of air in the growth of plants. One can hardly doubt, he asserted, that a great deal of air enters into the formation of vegetable substances and that it constitutes a considerable portion of their solid parts. As Hales and others have demonstrated,

> air exists in two modes in nature. Sometimes it appears as a highly attenuated, highly dilated, and highly elastic fluid, such as the one we breathe. At other times it is fixed in substances and combines intimately

with them, losing all its previous properties. Air in this state is no longer fluid but rather becomes solid, and it can only regain its fluidity if the substance with which it is combined is destroyed.[46]

Lavoisier was not inclined to commend the experimental efforts of his predecessors, but Hales was an exception; he had demonstrated the facts about the nature of fixed air beyond reasonable doubt. "His experiments are so persuasive that they cannot be questioned and they have been frequently repeated for everyone to see in the lectures of Mr Rouelle."

This excursion into the nature of fixed air was a side issue, however, and Lavoisier soon returned to the main question, which was the convertibility of water into earth. Hales's experiments were relevant because they demonstrated that water was not the only possible source of the weight gained by growing plants. Since Lavoisier's predecessors had not measured the amount of air that was fixed, he dismissed their claims as unfounded. "The experiments on vegetative assimilation of water prove nothing regarding the possibility of changing water into earth." He therefore abandoned this line of investigation and turned instead to experiments of a more chemical nature.

Chemists had long wondered whether water, if repeatedly distilled, could be converted into earth. To examine this possibility Lavoisier made use of a piece of laboratory apparatus called a pelican, a distilling flask designed so that the condensate drains back into the liquid being distilled. A carefully weighed sample of pure water was sealed in the apparatus and then repeatedly evaporated and condensed. In his second memoir on water, which was completed in May 1769, Lavoisier described in great detail how he conducted this experiment. He kept the pelican at a constant heat for 101 days between October 1768 and the following February and then analyzed the small amount of residue that had accumulated in the water.[47] Careful gravimetric analysis convinced him that this residue consisted of particles dissolved from the apparatus and "that the nature of water is not changed, nor does water acquire any new properties through repeated distillation."[48]

Lavoisier had the manuscript of this memoir initialed by the secretary of the Academy on 10 May 1769. Two months later he began a series of excursions into the provinces, some of them for geological mapping and others for business connected with tax farming. Although he returned to Paris between trips, he was not able to read his memoir on the conversion of water to the Academy until November 1770. He did not have much time for scientific investigations during the following year either, and he was no doubt considerably distracted towards the end of 1771 by preparations for his wedding in December. But during the early months of 1772, when he was once again able to give most of his attention to scientific studies, he soon got his research program back into high gear.

In February 1772 the Academy heard a report of a memoir by the prominent provincial chemist, Guyton de Morveau. The report was titled "On phlogiston considered as a substance having weight because of the changes in weight it produces in substances with which it combines"; the memoir itself was published toward the end of the year in Dijon.[49] After examining metals and other materials, Guyton concluded that the weight they appear to gain when calcined is caused by their loss of phlogiston. Lavoisier attended the meeting at which this report was read and Guyton's paper was evidently well received, for he was promptly elected a correspondent of the Academy.[50]

Lavoisier may have been impressed by Guyton's experiments and conclusions, but at the time he was preoccupied with a different line of investigation that he and several other academicians had just initiated.[51] Their plan was to destroy samples of diamond by heating them intensely, their purpose being to understand better how minerals are transformed into other states of matter. Lavoisier read an account of their investigations to the public meeting of the Academy in the spring of 1772.[52] Further experiments on the destruction of diamonds followed during the summer, and while reflecting on them Lavoisier again noted, but did not yet mention in public, his growing reservations about the concept of phlogiston.

There was also considerable discussion of fixed air in the Academy in the summer of 1772, the impetus being the arrival in Paris of the news that Joseph Priestley in England had succeeded in fixing air in water.[53] Priestley had devised a simple means for impregnating water with the air given off by the reaction between sulfuric acid and chalk. The Scottish chemist Joseph Black had described the chemical properties of this "fixed air" (carbon dioxide) in 1754 and many British chemists and physicians had come to believe that this, the first of the gases of the atmosphere to be subjected to chemical scrutiny, had beneficial medical properties. Priestley's *Directions for Impregnating Water with Fixed Air*, published in June 1772, was therefore primarily of interest to physicians and others concerned with preventing or treating certain types of diseases.

The fact that air had been fixed in a liquid was also of theoretical interest as well. Hales had long since taught investigators how to extract fixed air from solids and liquids, but they had not known how to imitate in the laboratory the processes by which nature fixes air in such bodies. Priestley's achievement was therefore considered significant. But the proud academicians of Paris were not entirely happy with this news and, to counter any suspicion that English chemists might be stealing a march on their French colleagues, they quickly pointed out that two decades earlier the French chemist Venel had made a similar gaseous water, which he called Selzer.[54] Nollet captured this French attitude towards British science nicely when he noted in 1760 that he seldom found scientific reports from England both novel and true.[55]

News of Priestley's discovery was conveyed by the "intelligencer" Jean Hyacinthe de Magellan. He drafted a long summary in French of Priestley's pamphlet and forwarded a copy, along with several copies of the original, to Trudaine de Montigny.[56] Trudaine, Director of the Royal Bureau of Commerce and a keen amateur scientist, had succeeded in 1764 to his father's position as an honorary member of the Academy. He had put together an impressive chemistry laboratory in his house and was eager to cultivate a close scientific relationship with Lavoisier. Trudaine therefore immediately forwarded Magellan's report and a copy of Priestley's pamphlet to Lavoisier and urged him to repeat Priestley's experiments.[57] Lavoisier thought well of Trudaine, who in time became his friend, patron, disciple, and to some extent collaborator, and on 18 July 1772 he read Magellan's letter to the Academy. A translation of Priestley's pamphlet was published the following month in François Rozier's journal *Observations and Memoirs on Physics, Natural History and the Arts* (*Observations et mémoires sur le physique, sur l'histoire naturelle et sur les arts*).

We do not know exactly how Lavoisier arrived at the realization that

2 An engraving of a large burning lens made for the Academy of Sciences. No picture survives of the instrument Lavoisier and his colleagues used in their experiments in 1772; the instrument shown was constructed a couple of years later.

many of the problems he was investigating involved the fixation and release of air. What we can say with some confidence is that no single piece of information or moment of revelation suddenly illuminated all these problems at a stroke. Throughout his career Lavoisier pressed forward along a broad research front.[58] In 1772 his larger goal was to associate changes of state that could be experimentally quantified with significant chemical changes. Guyton's work got him thinking about the weight changes that occur when metals are calcined, his collaborative investigations on the destruction of diamonds raised questions about the effects of fire on minerals, Priestley's experiments suggested new ways of studying the fixation of air, and the fixation of water in crystals provided a model of how fixation in general might alter physical and chemical properties. Lavoisier was vigorously pursuing a disparate and complex set of problems and he had no idea where on this broad front a significant breakthrough might occur. But he knew in principle what kind of methodological and theoretical resolution he was looking for and he seems never to have doubted that he would succeed in the task he had set himself.

As has already been mentioned, Lavoisier read a report at the 1772 Easter meeting on the recent experiments he and his colleagues had performed on the destruction of diamonds. Two months later the academicians Cadet and Brisson asked that the Academy's large focusing lens be brought out of storage and made available so that they could subject diamonds to a more intense heat than could be attained in their laboratory furnaces.[59] Their request was granted and Macquer and Lavoisier were instructed to work with them. Early in August, shortly before they began their experiments, Lavoisier composed a private note he titled "Reflections on experiments to be undertaken with the burning glass."[60] Because he believed certain ideas described in this note might be of considerable significance, he had it initialed by the secretary of the Academy on 8 August. This August memorandum tells us a great deal about what was on Lavoisier's mind at this crucial juncture in the evolution of his research program.

Lavoisier's delight in having access to the large lens was exactly like that of twentieth-century physicists who run experiments on gigantic particle accelerators – it was a machine capable of blasting apart substances previously considered immutable and thereby subjecting certain received notions about elementary matter to experimental testing. At the beginning of the century the academic chemists Homberg and Geoffroy had used the same Tschinhausen lens to study metals. Lavoisier had been struck by the similarity between the Stahlian theory of metals and the theory Geoffroy advanced to explain how the heat produced by the lens "reduced" metals to calxes.

Geoffroy argued that metals are composed of an earth and an inflammable principle or oil. This inflammable part, like phlogiston, can be separated and passed from one metal to another. It is present in animal and vegetable matter, as well as in minerals, and gives metals their distinctive luster and ductility. But Lavoisier was not fully persuaded, and he once again expressed in private his misgivings about the phlogiston theory: "It must be acknowledged that even today we know too little about the nature of what we call phlogiston to be able to say anything very precise about it."[61] Characteristically, he responded to the inadequacies of the available theories by proposing additional experiments. Geoffroy had used the lens to study metals; he would examine "an infinitude of mineral substances."

Experiments on the vaporization of liquids suggested how the burning lens could be used to bring the study of phlogiston under experimental scrutiny. By boiling liquids in evacuated chambers scientists had demonstrated that vaporization is caused by heating, not by the dissolving of liquids in air. When metals are heated over open fires, the vapors they give off are mingled with the fumes rising from the furnace. But if metals and minerals were placed in evacuated chambers and heated to a high temperature with the burning glass, the vapors given off could be collected in an unadulterated state. As was so often the case, Lavoisier was seeking to solve an important theoretical problem by finding a novel instrumental way to collect new evidence.

How diamonds are destroyed by intense heating was not at all obvious. Lavoisier thought they volatilized rather than burned; he also suspected they fell into a powder because they were suddenly exposed to the air after being heated. When metals are heated, he hypothesized, they give off a vapor that we should be able to capture in a properly designed apparatus. His thoughts on this question seem to have been guided by the similarities between the water of crystallization he had captured in his earliest experiments on gypsum and whatever it was that metals give off when they are heated. The fixation of elements and the formation of solid substances remained firmly linked in his mind. As for the minerals, he considered them insoluble salts composed of acids and bases. The chemist's task was to decompose them through either intense heating or the use of chemical reactions. The fixation of light was another problem Lavoisier hoped to address with the burning mirror, his particular concern being to explain why a flame, but not concentrated sunlight, will ignite inflammable substances.

The last section of the August memorandum is "On fixed air, or rather, on the air contained in bodies." Here Lavoisier revealed that he believed it was fixed air, and not an oil or phlogiston, that combined with minerals to form metals and give them their distinctive properties. Effervescence is simply the release of the air fixed in the bodies that are reacting; he said nothing more about it because he planned to address the topic again in a work he was then revising. He noted, however, that most metals, after being heated for a long time in the focus of the burning lens, do not effervesce in reactions and are nearly insoluble in acids. "No doubt," he concluded, "the degree of heat to which they were exposed released the air with which they were combined."[62] He realized these observations and speculations required further experimental confirmation; he was especially eager to use Hales's collecting apparatus "to measure the quantity of air produced or absorbed in each operation." As the August memorandum reveals, Lavoisier was increasingly concerned with the fixation and release of air, but his working hypotheses were still quite speculative and his experimental program still lacked a sharp focus.

The possibility of being anticipated haunted Lavoisier during the remainder of 1772. At a meeting of the Academy held a few days after his August memorandum was initialed, the secretary read a paper on the evaporative cooling of liquids that Nollet had placed on deposit in the mid-1750s. News had recently reached Paris that Joseph Black in Edinburgh had formulated a coherent and quantified theory of latent heat. The academicians responded by exhuming Nollet's paper from the files to establish French priority in this area of research.

During the discussion period following the reading of Nollet's paper, Lavoisier hurried out, gathered together some manuscripts, and returned with a draft "Memoir on elementary fire" that he asked the secretary to initial. He did not disclose the contents of this bundle, but later on he identified it as part of a "System of the elements" he had been working on.[63] We now know that Lavoisier considerably overstated the degree to which he had mastered the problems addressed by Black's theory. The notes he described as a system of elements were as tentative as his August memorandum, yet they do indicate quite clearly that on the vaporization of water he remained a devoted follower of Nollet and that he was struggling to broaden that theory to include the fixation and release of air and fire as well.[64]

Evaporation causes cooling, Lavoisier wrote, because "evaporation is entirely a combination of any sort of matter with substantial fire." The vapor formed by water and fire is practically identical to air itself.

> Water combined with fire forms a new fluid very analogous to air . . . It appears that the combination of water and fire endures in a vacuum and that this vaporous fluid behaves exactly like air. It appears to be equally certain that if the earth's aerial atmosphere was destroyed another analogous one composed of vaporized water would form.

These assertions allowed Lavoisier to glide effortlessly from the problem of the fixation of air to that of the fixation of specific vapors. Fixed fire was the key to his "singular theory," which was based on the assumption "that the air we breathe is not a simple substance; it is a specific fluid combined with the matter of fire." Although he did not yet know how to put it all together, Lavoisier was determined to formulate a coherent and symmetrical theory for the fixation and release of the elements air, water, and fire: "What we will say concerning air is equally true of phlogiston or the matter of fire."

At the end of August Macquer was promoted to the rank of pensioner in the academy and Lavoisier was promoted to associate. In September Lavoisier began a series of experiments on a sample of German phosphorus he had purchased from the pharmacist P. F. Mitouard.[65] Mitouard had been one of the collaborators on the diamond experiments and Lavoisier and Macquer were actively promoting his candidacy for election to the Academy. On his own Mitouard had found that when phosphorus is burned and the vapors are condensed to form a liquid acid, the weight of the acid is always greater than the original weight of the phosphorus. What remained uncertain was whether the added weight came from water vapor in the air or the fixation of air itself. Lavoisier knew of this finding and considered it most remarkable. He set out to investigate this matter further, but his initial efforts were inconclusive.

In October Lavoisier performed several experiments on phosphorus. Unable to make any headway in clarifying the source of the weight gain but convinced that phosphorus, like many other minerals, fixes air, he began a systematic investigation of the various salts that are formed when its acid combines with earths, alkalis, and metals. He devoted most of the month to this traditional form of chemical investigation but terminated this line of research before completing the paper in which he planned to summarize his findings. Once again he asked the Academy's secretary to initial a draft so that his interest in a project would be protected while he rushed off to engage another, suddenly more urgent question.

The investigative breakthrough Lavoisier had been looking for occurred during the last ten days of October 1772. Evidently he happened to think of a simple way to separate the weight phosphorus gains during combustion from the weight of the water that combines with its vapors during the formation of phosphoric acid. In retrospect it is easy to see the similarity between Lavoisier's earlier hydrometer experiments on relative densities and his resolution of this problem, but had the similarity been obvious to him, he would have acted sooner. Lavoisier described his method in detail in a hastily prepared outline of the paper he hoped to read at the next public meeting of the academy. As he was not able to secure a place on the schedule for the fall meeting, he had to wait for the spring meeting to lay his findings before his colleagues and the larger public interested in science.[66]

Lavoisier began his momentous experiment by preparing phosphoric acid in the standard way. He burned carefully weighed bits of phosphorus in a bell jar that had been rinsed in water and collected the acid that formed when the vapors condensed. He then poured the acid into a beaker and marked and weighed it. The acid was then poured out and the beaker was filled to the mark with distilled water and weighed. The difference in the weights of the two equal volumes of liquid was a measure of the weight of the acid in the liquid. An additional calculation told him how much weight the phosphorus had gained during combustion; it was considerable. He found that the weight gained in combustion was more than a third of the original weight of the phosphorus. Far from being puzzled, Lavoisier was exultant. He had found a technique that told him how to choose among many of the theoretical possibilities he had been considering and how to transform his speculations about the elements into demonstrated knowledge. Unsolved problems remained, of course, but he was convinced he had made a revolutionary breakthrough.

The next step was to use this method to determine the weight gained by a much more extensively studied mineral, namely sulfur, when it is burned. Lavoisier reported that he performed the experiment immediately and again found that the weight of the sulfuric acid he collected was much greater than the weight of the original sulfur.[67] This experiment appeared to confirm that he had made a chemical discovery of great significance, for the properties of sulfur were of central importance to the phlogiston theory. Lavoisier quickly sketched out the theoretical implications of his discovery. His main claim was that he could now explain the weight gain that occurs during calcination.

On 1 November Lavoisier drafted and sealed a brief note for deposit at the Academy; he labeled it "On the cause of the weight gained by metals and several other substances when they are calcined."[68] Seeing combustion and calcination as essentially the same processes carried on at different rates was not in itself novel; what was novel was the ability to explain the well-known but puzzling weight gain in calcination. Lavoisier also demonstrated that he could explain other instances of calcination by, for instance, reducing litharge, the calx of lead, when heated with charcoal in a pneumatic collecting apparatus of the sort Hales had used. Because this was essentially a demonstration experiment designed to verify his by then firm conviction that the gain in weight in both combustion and calcination is caused by the fixation of air, he simply noted the great volume of air given off by litharge and made no attempt to weigh it. The important point, as he said in his sealed note, was that he had made a discovery that was "one of the most interesting to be made since Stahl."

As he drafted the report he would read to his colleagues in the Academy and to the public, Lavoisier explored the larger significance of his discovery. The generalizations that occurred to him were not, of course, inherent in the experiments and observations themselves. The meaning of his experiments became apparent only as he integrated their results into the patterns of theoretical explanation that informed his entire research program. Air had emerged as the key element, and having found this key, Lavoisier suddenly acquired an enormously increased respect for Hales's investigations of fixed air. Since he could not describe all the experiments he had already performed, Lavoisier wrote, "I will only say that while reviewing all of Hales's reports and repeating his experiments, I realized beyond any doubt that the increase in weight occurs because a portion of air is absorbed and fixed along with acid vapors."[69] He concluded that since the weight gain in the formation of acids is similar to that in the calcination of metals, we may presume they have the same cause.

Lavoisier was unsure whether his discovery would lead to the refinement of the phlogiston theory or its replacement.[70] The draft report he composed during the closing weeks of 1772 addressed this theoretical problem and several others as well.

> It is evident that Stahl's theory on the calcination and reduction of metals is seriously flawed and needs to be modified. It treats calcination solely as a loss of phlogiston, although it has been demonstrated that a loss of phlogiston and an absorption of air occur at the same time ... One can no longer doubt that air is materially combined in metallic calxes and that this agent combines according to the increase in weight observed ... Finally, it is obvious that air is the acidic substance of Mr Meyer in most if not in all cases and this proves that fixed air often functions as an acid in nature.[71]

Lavoisier had no way of knowing at the outset how far he could ride this new horse, but he was convinced he had made a breakthrough of great importance. He believed he had good reason to concentrate on the chemical effects of fixed air and he was confident that he commanded the experimental techniques and skill needed to do so. His chemical apprenticeship was over; he had embarked on the creation of his masterpiece.

In February 1773, several months before reading his epochal memoir at the Easter meeting of the Academy, Lavoisier summarized his new research program in a note he wrote on the first pages of a new laboratory notebook.[72] This private memo was first published nearly a century after Lavoisier's death, and because it contains what was then thought to be his earliest prediction that his research would bring about "a revolution in physics and chemistry," it has long been considered a uniquely prescient document in the history of science. But when seen in the context of the story told above, this note appears to represent little more than an incremental step forward in an evolving research program. Lavoisier was still struggling to gain control over his new subject. He therefore committed himself to surveying all the relevant literature on fixed air, to examining critically all previous theories proposed to explain its effects, and to repeating all the experiments of those who had studied the subject. His ambition, confidence, energy, and determination were in full cry:

> I have felt no obligation to consider anything done before me as more than a hint; I propose to repeat everything with new safeguards so as to join what we know about air that is fixed, or is released from substances, with knowledge acquired elsewhere so as to formulate a theory ... A vast number of additional experiments must be performed, however, before the whole can be grasped.

The narrative account given above provides a coherent historical explanation of how Lavoisier progressed from the research problems to which he was introduced by his masters to focusing on the specific issues that led to the theories for which he is remembered. Yet we should not forget that it is our knowledge of where this line of research led, knowledge that was unavailable to Lavoisier, that gives this narrative direction and meaning. Like all stories, this account of Lavoisier's research is highly selective. To help balance its orientation toward an achievement that lay over the horizon, we will end this chapter with brief descriptions of two contemporary documents he wrote that did not contribute to his understanding of pneumatic chemistry. The first of these, a short paper on the crystallization of salts read to the Academy toward the end of 1772, will serve to remind us that, even while working intensely on the research frontier, Lavoisier continued to be actively involved in the day-to-day political life of science. The second document, an excerpt from the preface of Lavoisier's first book, provides a concise summary of the broad range of problems that continued to interest him even while he was struggling to sort out the mysteries of fixed air.

The short memoir on the crystallization of salts that Lavoisier read to the Academy on 25 November 1772 is both an example of his dazzling rhetorical skill and a reminder of how difficult it is to recover the full significance of historic texts. The memoir begins with two concise paragraphs on methodology which can still be read with profit.

> No matter how much at odds with received and accredited ideas a fact may be, no matter how much it appears to contradict the laws of physics, it is only with the greatest caution that one should say it falsifies (*nier*) them. Simple assertions are unconvincing and a physicist should not contest established beliefs unless he is well armed on all fronts with arguments, instruments, and experiments.
>
> The caution that physicists should exhibit in suspending judgment on falsification is of even greater importance with regard to verification (*affirmer*). They should accept new discoveries only after careful critical review and their critique should be especially severe when the new facts cannot readily be brought into agreement with truths that are thought to be well grounded.[73]

Guided by these cautionary thoughts, Lavoisier turned to the examination of a brief account of the crystallization of salts by the well-known pharmacist, Antoine Baumé. Baumé's report had recently been published in the reputable weekly journal *The Harbinger* (*L'Avant-Coureur*).[74] One reads there, Lavoisier noted, that the effects of attraction and repulsion on crystallization can be detected at quite a distance. If it can be demonstrated, as is claimed, that this power is effective at distances of more than a foot, then this is truly one of the most singular phenomena in physics and chemistry. Lavoisier therefore decided to repeat Baumé's experiments and provide a public report of his findings.

The experiments themselves are, by twentieth-century standards, as misguided as eighteenth-century attempts to weigh fire, but Lavoisier never smiled. He made up a saturated solution of Glauber's salt (hydrated sodium sulphate) and poured it into several beakers. After placing these on separate tables, he set near each of them other beakers containing crystallized Glauber's salt, solid fixed alkali (K_2CO_3), saturated solution of fixed alkali, and so forth. He reported that he exercised great care and repeated the experiments several times, but in no case did he find any difference in the crystals of Glauber's salt that formed in the beakers containing the saturated solution. Reading this, one is reminded that at a later date Lavoisier was obliged to assess carefully the claim that underground water can be located with a divining rod.

Having disposed of Baumé's claims, Lavoisier took the occasion to report on an interesting experiment of his own. He put a thermometer in a warm saturated solution of Glauber's salt. The thermometer reading fell as the solution cooled, but, when the first crystals began to form, the thermometer reading stopped falling and after a while rose slightly. It began falling again only when all the salt had crystallized. "This experiment proves," Lavoisier incautiously asserted, "that the cooling that accompanies the solution of Glauber's salt in water is comparable to the heat that appears during the crystallization of the salt." Furthermore, since these and related phenomena fall under "the laws of physics and are all the result of the same cause, a physicist can easily explain them."[75]

Viewed historically, this memoir illustrates in miniature the cut and thrust of eighteenth-century scientific controversy. Lavoisier knew full well he was not the first to observe the heat given off during crystallization and hence he presented his experiment as a mere afterthought. One of his purposes, of course, was to demonstrate yet again that Joseph Black had not discovered latent heat. Thus three days later, when Macquer rebuked Lavoisier publicly by pointing out that Baumé had fully described the heat of crystallization in his soon to be published three-volume *Experimental and Rational Chemistry* (*Chymie expérimentale et raisonnée*), Lavoisier could let it pass without comment.[76] His carefully constructed report had already scored points for Nollet and the experimentalists in their continuing battle with the Buffonian Newtonians. Baumé had merely provided Lavoisier with a convenient target of opportunity.

But there was a longer history that this engagement made reference to as well. The Scottish physician and chemist William Cullen, who was Black's mentor, is credited with being the first to describe in a general manner the cooling effects of evaporation, but his paper, published in

1756, was not widely known in France.[77] In 1757 Baumé read a paper on this subject to the Academy of Sciences, of which he was not a member, and then worried throughout the following decade, as his paper remained unpublished, that his investigation would not receive the notice it deserved. His fears were not groundless, for the reduction of temperature during certain changes of state had already attracted the attention of many investigators, most of whom hoped to develop practical methods of cooling that could be used in hot climates. In 1756 Nollet had suggested replacing ice with the cooling that occurs when saltpeter or sal ammoniac is dissolved in water; the next year he proposed further study of the cooling effects of evaporation that Baumé had described to the Academy. He also gave an account of his own investigations of these subjects in a sealed note that was not opened until 19 August 1772, when the Academy first heard of Joseph Black's work on this subject. As noted above, it was the reading of Nollet's note that prompted Lavoisier to rush out for his "System of the elements" and have it initialed.

But why did Lavoisier devote time during his frantically busy schedule at the end of 1772 to refuting Baumé's unlikely claims about the effects of attraction and repulsion on the formation of crystals? Perhaps it was because he recognized a moment of vulnerability on the part of his opponents and wished to exploit it before it passed. In a book published in 1758, a copy of which Lavoisier owned,[78] Baumé had been seriously if unfairly charged with stealing Cullen's ideas on the cooling that accompanies evaporation. But Lavoisier would not have harried Baumé for this reason alone had he not had bigger game in sight.

Baumé was Macquer's collaborator and protégé and had long sought an appointment to the chemistry section of the Academy. Baumé and Macquer were Buffonians who used attractionist concepts to explain chemical phenomena. Lavoisier had not forgotten that Buffon had championed the successful candidacy of Jars in 1768, when Lavoisier had been appointed as a supernumerary. Both of Lavoisier's masters, Nollet and Rouelle, had been interested in salts, and Lavoisier had committed himself to using Nollet's physicalist theory of heat to make sense of the chemical properties of salts described by Rouelle. The next election to the chemistry section would take place in December, when Baumé was in fact finally elected. Thus in late November Lavoisier had many interconnected reasons for taking Baumé to task for his insupportable attractionist theory of crystallization.

Such an elaborate reading of so slight a text may seem rather labored, but the context in which this work was composed and presented was far from simple. Lavoisier was a brilliant tactician who knew how to marshall his forces across a broad front; he also knew how to employ masks and indirection to avoid getting enmeshed in draining personal controversies. He could not predict how each scientific engagement would turn out or where each purported breakthrough would lead, and he therefore had to be constantly campaigning for the individuals and ideas he championed.

Declaring that he did not wish to be accused of partiality, Lavoisier ended the first part of his 1774 *Physical and Chemical Essays* with a long excerpt from Baumé's recent chemistry text; the passage he quoted was in fact a clumsy attempt to refute those seeking to effect "a complete revolution in chemistry" by replacing phlogiston with fixed air.[79] And over a decade later, in his 1785 "Reflections on phlogiston," Lavoisier focused his rhetorical attack on the theories of Macquer and Baumé. Clearly Lavoisier considered Baumé a good enough chemist to serve as a whipping boy for the phlogiston theory. Being treated in this manner gave Baumé more than enough reasons to oppose Lavoisier's new chemistry, and he persisted in his defense of the phlogiston theory until the end of the century. Thus while Lavoisier's 1772 critique of Baumé's crystallization experiments is irrelevant to the internal development of pneumatic chemistry, it does serve as a reminder that the serene ambience of fraternity that characterized the public face of eighteenth-century science was in fact a mask that hid the intense interactions that heated its core. Lavoisier excelled at the courtly competition of reason and rhetoric and played the game of science with vigor and great panache. Above all, he played to win.

Lavoisier spent most of the year 1773 mastering the subject of fixed air. The results of this effort were collected and published early the following year in his first book, *Physical and Chemical Essays*. This volume provides ample evidence of the effort and intelligence he devoted to the task. But it should also be noted that, while focusing his attention on the study of fixed air, Lavoisier avoided limiting himself to that topic alone. As he indicated in the preface to his *Essays*, once he had begun investigating a subject he was loath to let go of it. He had made one major scientific breakthrough and expected to make others. He was still a young man and at that point had no idea how long it would take him to complete the revolution he had just begun.

This first volume will, I hope, be followed by several others, and in these I shall pursue a number of experimental programs that I have already begun and wish to continue:

1 On the presence of the same elastic fluid in a great many natural bodies in which its presence has not been suspected;
2 On the complete decomposition of the three mineral acids;
3 On the boiling of fluids in the vacuum of the air pump;
4 On a method for determining the amount of salt contained in mineral waters according to their specific weights;
5 On using a mixture of alcohol and water in the analysis of complex mineral waters;
6 On the cause of the chilling that accompanies the evaporation of fluids;
7 On certain optical questions that I investigated while preparing a memoir on lighting the streets of Paris, a work the Academy honored at its public Easter meeting in 1766 with a gold medal and that I have revised and expanded considerably;
8 On the height of the main hills near Paris in relation to the level of the Seine river, as measured with a good quadrant circle belonging to the chevalier de Borda . . . Lastly I will include a list of many barometrical observations made in the different provinces of France. I will also include a cross-sectional map of the land in these provinces that goes to a considerable depth, gives the order in which the strata occur, the constant level at which certain substances (such as shells) can be found, and the remarkable inclinations that some formations always exhibit.

These various works are mostly well advanced and many were initialed long ago by Mr de Fouchy, permanent secretary of the Academy. I hope I shall soon be able to submit them to the judgment of the public.[80]

5

The Company of Tax Farmers

The story of Lavoisier's marriage is pure eighteenth-century theater. All the classic roles are represented: a powerful patron, a manipulative broker, a virtuous maiden, a distraught father, an innocent gallant, a chorus of relatives. Anxiety and delight alternate as the plot progresses from impending calamity to joyous resolution. The action culminates in a brilliant finale in which all parties are reconciled. With very little elaboration it could serve as a libretto for Mozart or Gilbert and Sullivan.[1]

If Lavoisier gave any thought to marriage before November 1771, he did not allow it to distract him from the many activities to which he had committed himself. In March 1768, just a few months before his election to the Academy, he decided, on the advice of a family friend, to purchase a share in the Company of General Farmers (*fermiers-généraux*). This Company was a financial partnership that underwrote and managed the collecting of certain indirect taxes. Its members were called farmers because they operated under terms stipulated in a lease (one meaning of the term *ferme*), not because they were in any way involved in agriculture. The taxes they were responsible for were actually collected by a vast army of agents employed by a branch of the civil

110

administration called the Royal General Farms (*ferme générale*). The operation of these institutions and Lavoisier's involvement with them will be described in greater detail later in this chapter.

What concerns us here is that, having joined the Company of General Farmers, Lavoisier was obliged to undertake several extended tours to inspect and report on the agents of the General Farms. These trips through the provinces, along with his mapping project with Guettard, kept him away from Paris for much of 1769 and 1770. On matters of concern to the Company Lavoisier reported to a senior partner named Jacques Paulze.

Paulze had married the niece of the abbé Terray, who was then a lowly law clerk, in 1752. Paulze's wife died after bearing him three sons and a daughter, Marie Anne. In 1771 the girl was thirteen years old and about to leave the convent in which she had been reared. Terray had by then become the Controller-General of Finances, a post which included supervision of the Company of General Farmers; he was, in other words, a powerful politician and Paulze's boss. The Baroness de La Garde, who exercised considerable influence over Terray, had a fifty-year-old brother, the Count of Amerval. Amerval, being financially pinched, was looking to make a good marriage. The Baroness suggested to Terray that Amerval marry Paulze's daughter and Terray approved of the match.

What was poor Paulze to do? Although his daughter was still a child, she certainly knew that eventually a marriage would be arranged for her, yet she had spirit and was unwilling to accept this premature and unsavory proposal. She called Amerval "a fool, an unfeeling rustic and an ogre," words that must have torn at her widowed father's heart. At first he lay low, hoping the threat would pass, but the Baroness and Terray insisted. In desperation he sent Terray a respectful plea for reconsideration.

> My dear uncle, when you spoke to me of my daughter's marriage, I thought of this as an event that was far in the future and that there would be some similarity in age, character, fortune and so forth; this, I find, is not to be. Mr Amerval is fifty, my daughter is only thirteen; he has an income of only 1,500 francs, my daughter, while not wealthy, can now bring more than twice that to her husband; you are unfamiliar with his character, but I am reliably informed that it is hardly suitable to my

daughter, to you, or to me. My daughter does not like him; I will not have it done against her will.

The Baroness tried to sway Marie Anne with accounts of the brilliant presentation she would receive at court, but to no avail. Terray threatened to remove Paulze from his post as a director of the Company, but the other tax farmers protested that he was the only one who could put the affairs of the tobacco department in order. The situation grew increasingly dire, the dilemma ever more distressing.

It occurred to Paulze that the siege might be lifted if he were quickly to arrange a more suitable match for his daughter. At once he thought of Lavoisier. The possibility was broached to the young man, doubtless with the greatest delicacy, and the pact was hastily sealed. We have no record of Lavoisier's reaction; he can hardly be said to have courted his intended. In November it was agreed they would be wed in December; congratulations poured in from concerned friends and relations. As an aunt wrote to the relieved father,

> what joy that my niece has escaped the danger that surrounded her and is now about to enter into an arrangement in which she and you will find all the advantages and promise of the greatest happiness. She is so well reared and so reasonable that I have no doubt she will make her husband very happy.

But what of Terray? Would he act vindictively? Would the drama of sex and power be played out with the polite malice so charmingly chronicled in that aristocratic fantasy *Dangerous Liaisons*? The bourgeoisie had good reason to avoid such destructive conduct and on this occasion everyone behaved honorably. Terray's family urged him to accept the inevitable, and when presented with a *fait accompli*, he acceded without recrimination. He capped this conversion to virtue by arranging to have the couple married in his private chapel. Thus were the last clouds dispelled and the stage set for the grand finale.

The union was celebrated at the signing of the contract. A reception attended by more than 200 distinguished guests was held in the mansion of Terray de Rozières, brother of the Controller-General. Many of Lavoisier's patrons and friends were there, including Bertin, who had authorized the geological survey, and Sartine, who had initiated the prize for street lighting. Prominent institutions were represented.

The Academy of Sciences sent d'Alembert, Cassini de Thury, and Lavoisier's old teacher, Bernard de Jussieu; Guettard was out of town but conveyed greetings. Many members of the Company of General Farmers attended. Notable in their absence were the magistrates of the Parlement of Paris; they had been exiled by Maupeou at the beginning of the year for refusing to register edicts issued by the king. The wedding itself was held on 16 December 1771 in Terray's chapel. The bride was not yet fourteen, the groom was twenty-eight. Marie Paulze's two uncles, Terray and Terray de Rozières, attended as her witnesses.

This sentimental comedy of manners did not exhaust the meaning of the event. The conjoining of two families also involved financial arrangements that needed to be stipulated in detail. Paulze gave his daughter a dowry of 80,000 livres, 21,000 of which she received directly, the remainder to be paid over the next six years. Lavoisier brought far greater wealth to the union. He had already received 170,000 livres from his mother's estate and his father had given him a 250,000 livres advance on his inheritance. With these funds he borrowed nearly a million livres, from which he advanced 780,000 to the Company of General Farmers. This investment increased his stake in the Company from one-third to one-half of a membership in 1771. Taking on such a loan was standard practice for members of the Company, but doing so put Lavoisier at some risk and left him with heavy interest payments.[2] He must have worried that he might, like his father-in-law, lose money during his early years in the Company, yet after deducting his costs from the income he expected, he calculated he would realize about 20,000 livres a year from the Company. Lavoisier could also look forward to inheriting 50,000 livres upon the death of his aunt Constance, who had reared him following his mother's death, and a share of the estate of his great-aunt Lalaure. Lavoisier's father added yet more to this cascade of beneficence by buying a house for the bride and groom. He also purchased a royal office for his son, a position as King's Secretary (secrétaire du roi), that brought with it a hereditary title to nobility.[3] Although Lavoisier doubtless appreciated his father's intentions, he wisely avoided presenting himself as a privileged member of society and almost never used his aristocratic title.

Lavoisier's wedding can also be seen as completing his transition from the world of the parlementaires to that of the financiers. There was a considerable overlap between these two groups, yet there were also significant differences. Maupeou's recent attack on the Parlement of Paris appeared to indicate that the king would succeed in vanquishing the *parlementaires* and establishing his dominance over the administration of justice throughout France. The financiers, on the other hand, and especially those who were members of the prestigious Company of General Farmers, had established close working relationships with the king's ministers, and so long as royal finances depended on their services, they could look forward to ever greater profits and influence. Lavoisier had been encouraged to join the Company by the royal governor (*intendant*) of the province of Lorraine, not by one of his family's parlementary friends. In joining the Company, as in joining the Academy of Sciences, Lavoisier was openly allying himself with the Bourbon program for increased administrative centralization. Doing so obliged him to subordinate whatever attachment he may have felt to the parlementary ideology of decentralized authority based on traditional rights and privileges. Lavoisier chose to make his career in science and finance at least in part because he believed the nation could best be served through the advancement of reason and centralized public administration. Paulze had read his younger colleague's intentions and expectations correctly; he had chosen well in his daughter's behalf.

The more intimate aspects of this marriage also deserve a moment's notice, although neither Lavoisier nor his wife was inclined to display such matters in public.[4] The couple's respect and affection for one another were frequently remarked. Soon after their wedding Marie Anne began taking an interest in her husband's scientific research and she quickly mastered skills that enabled her to contribute significantly to what was often a joint enterprise. There are indications that she was pregnant in 1774, but no infant ensued and to their regret the marriage remained childless. Undaunted, she joined her husband in his laboratory, where he spent so much of his time. In 1775, when Marie Anne was only seventeen years old, Lavoisier received a letter in which Magellan asked after his "philosophical wife," and two years later it appears Marie Anne was being tutored in chemistry by her husband's disciple and collaborator, Jean Baptiste Bucquet.[5] She learned English, which Lavoisier never mastered, and translated many chemical works for him. She also studied drawing with the painter, Jacques Louis David, made sketches of Lavoisier's laboratory, and prepared plates for his *Elementary Treatise on Chemistry*. She frequently assisted in the laboratory, often recording in bound registers observations dictated to her, and she regularly presided over philosophical soirées after she and her husband had moved to their apartment in the Arsenal. Theirs was a marriage marked by genuine mutual affection and extraordinary intellectual companionship.

Traditional accounts of French royal finances in the eighteenth century treat the collapse of the old regime as foreordained. Revolutionaries quickly learned to vilify financiers as selfish manipulators who reaped hugh profits and lived lives of indecent extravagance while burdening the nation with unmanageable debt and extorting ruinous tax payments from an impoverished peasantry. Critics from across the Channel insisted that financial ruin was unavoidable so long as France failed to bring all aspects of public finance under the control of administrators answerable to a representative government. It is widely understood that exemptions from taxation, which the aristocracy and Church defended as traditional privileges, forced the poorest part of the population to bear all the costs of a luxurious and incompetent government. The state evidently lay helpless while financiers, who provided essential fiscal and monetary services, misappropriated the nation's wealth. The only interesting question appears to have been not why the old regime would fail, but when it would do so.

That the collapse of the old regime was signaled by the bankruptcy of the monarchy is not in dispute. According to recent scholarship, however, this financial collapse was caused more by political inaction than by inadequacies in the system established to handle financial affairs. From an economic and financial point of view, eighteenth-century France was a wealthy and relatively well-ordered nation. Administratively, the collapse of the old regime was not inevitable. Its bureaucracy was not unusually or insupportably corrupt, its ministers and advisors were reasonably competent and well-informed, and there was no reason to think its financial institutions could not be reformed to meet changing circumstances. The Revolution did not begin with a financial crisis, but rather with a political impasse that defied all attempts at resolution. When that political impasse forced a financial crisis, the extent and seriousness of the government's failure were made obvious to all. It was at that point that political revolution became a real possibility.[6]

Lavoisier joined the Company of General Farmers when he was twenty-five years old. His decision to become a tax farmer should not be taken as evidence that he was cynical, callous, or avaricious. The reality of the situation was more mundane. In joining the Company Lavoisier was investing in a thoroughly legitimate enterprise, one which, like the modern bond market, provided the government with an essential service. The Company, like all large complex organizations, had its share of problems, yet Lavoisier evidently believed he could address whatever injustices and inefficiencies he encountered while working within it. He had no reason to think the Company was radically corrupt or corrupting, nor that it was bound to fail. Quite the opposite was true. When seen as part of the long campaign for monarchal centralization, the Company of General Farmers itself appears to have been a highly progressive institution. Among Lavoisier's several reasons for joining it, surely one was a desire to contribute to the welfare of the nation by helping to make this leading financial institution more responsive and efficient.

Lavoisier spent his first few years in the Company learning the ropes and developing his managerial skills. His responsibilities were administrative rather than those of a policy analyst or political adviser. Given the constraints of the multi-year leases under which the Company operated and the difficulties involved in collecting the taxes assigned to it, the working partners faced daunting managerial problems. In the mid-1770s, when he became a prominent figure in the Turgot ministry, Lavoisier took on more extensive administrative responsibilities within the Company and spoke more openly on economic issues. In addition to arranging loans to the government and overseeing the collection of taxes to repay them, he worried over how the wealth of France might be increased, how the nation's tax base might be broadened, and how the rapid growth of royal indebtedness might be checked. His activities in these later years and his views on these broader concerns will be discussed in chapter 8. In this chapter we will concentrate on describing the organization and purpose of the Company itself and Lavoisier's activities during his early years as a member.[7]

The collecting of taxes had long been leased out in France. It began locally, long before the establishment of modern monarchies and before medieval duties were converted into taxes paid in cash. As the French state grew, the number of taxes and tax leases increased. Political authorities having the power to tax found the system highly attractive. Collecting taxes was not, however, the main function of those who entered into tax leases with the government; that was merely a way of paying for services rendered. The system's main purpose was, like the modern bond market, to give political leaders access to financial credit. Financiers who entered into leases with the government agreed to provide lump sum payments upon signing and further payments at set intervals during the life of the lease. They then collected certain taxes to pay the costs incurred in raising these funds and to make a reasonable profit.

When new leases were being negotiated, both the government and the financiers naturally sought favorable terms. Since the tax yield over a number of years was not guaranteed, the financiers took real risks and occasionally suffered losses. Loans made to powerful political entities could also be repudiated or, more drastically, the creditor and his property could be seized. Yet when viewed from the perspective of the late twentieth century, another era of enormous public indebtedness, the system of tax leasing appears to have had some real advantages. Privatizing governmental financing should have made the state more sensitive to market-based estimates of its credit worthiness, restricted the growth of public-sector employment, and dispersed the administrative exercise of state power. In principle at least, leasing out tax collection seemed a perfectly reasonable way to meet the French state's growing financial needs.

Colbert, Louis XIV's great minister, recognized the advantages of tax leasing and made them a part of his absolutist program for administrative centralization. In 1681 he completed a process that had been begun many years earlier by consolidating control over nearly all indirect taxation and assigning it to a single organization called the Royal General Farms. This agency was given responsibility for a total of 137 indirect taxes of four different sorts: the *gabelles*, taxes on salt; the *traites*, customs duties within France; the *aides*, excise and sales taxes; and the *domaines*, revenues collected from royal properties. The General Farms, a branch of the royal civil administration, had a staff of roughly 30,000, making it the largest employer in France after the royal army and navy. Its 21,000 uniformed agents were armed and had legal power to enter and search households and seize property. In addition to collecting and forwarding taxes, the agency was responsible for the production, distribution and sale of certain state monopolies; these included salt and, after 1730, tobacco.

Tax collectors are universally viewed with suspicion, but the operating practices of the General Farms were especially aggravating. Colbert's reforms centralized the collection of indirect taxes, yet the structure of taxation itself was not rationalized. Systematic collection increased the yield from existing indirect taxes, and the royal income from the General Farms increased dramatically during the eighteenth century. The same taxes that provided the government with 99,000,000 livres in 1725 yielded over 253,000,000 in 1788, an increase of nearly two and one-half times, while during the same period receipts from direct taxation, which was not leased out, roughly doubled, going from 87,500,000 livres to somewhat over 179,000,000. At the local level, however, this increased efficiency in collection severely exacerbated the grievances generated by an allocation of tax burdens that was undeniably irrational and inequitable.

Most tax obligations in the eighteenth century were legitimated by tradition, but to the individual payer it appeared that agreements sanctioned by custom were being violated when the cost of compliance suddenly increased. Because tax collection was rationalized long before any real progress was made in rationalizing tax burdens, the political legitimacy of tax collection eroded rapidly during the middle decades of the century. Vigorous collection encouraged and appeared to justify deception and smuggling; the increased use of these techniques of avoidance led to more intrusive policing and yet greater resentment.

It was not the centralization and rationalization of administration that made the General Farms so hated, but rather the partial and politically traumatizing manner in which that reform was carried out. The General Farms were pressed by the financial needs of the government and constrained by royal deference to traditional privileges and exemptions. It is thus hardly surprising that the General Farms' structural problems, when viewed in retrospect, have often been called insoluble. Yet in the eighteenth century it was not unreasonable to believe that indirect taxation could be made part of a thoroughly reformed, administratively rational, and politically legitimate national tax system. But as everyone knew, many political eggs would have to be broken to make this particular omelet.

The government had two ways of managing the Royal General Farms. It could either create a specific administration (*régie*) staffed by civil servants with immediate responsibility for collecting indirect taxes, or it could lease out this managerial responsibility to a group of investors. Both methods were used in the eighteenth century. Putting the agency under the direct control of a minister was a way of eliminating the problem of dealing with financiers who sought favorable terms from the government. The crucial difference, however, was that the leasing arrangement provided the government with immediate credit, something salaried ministers could not provide. Had the royal government lived within its income, or had it had access to credit through a national bank, it could have dispensed with the services of private financiers. In fact, however, neither of these possibilities was realized. Royal ministers were therefore obliged to rely on private financiers to service the nation's large and growing debt. One way to obtain their assistance was to grant them negotiated leases to manage the General Farms.

From 1726 until 1789 the management of the General Farms was in the hands of the Company of General Farmers. The Company, essentially a multiple partnership of forty to sixty members, signed a series of six-year leases, each of which was named for the leading partner; thus, for example, the lease negotiated in 1768, when Lavoisier joined the Company, was known as the Alaterre lease. When a lease was negotiated, the Company agreed to give the government a fixed amount of money over the course of the lease and provide it with an immediate cash advance of from 10 to 50 percent.

Administration of the General Farms was handled by a series of committees whose members were appointed by the Controller-General. Every member of the Company received three types of income: an annual managerial salary, interest paid on capital advanced by the Company, and a share in the profits realized from the excess of tax collections over operating expenses. As the government's needs and indebtedness increased, the Company was pressed to increase the lease price at each renegotiation and to provide more short- and long-term credit. The Company resisted, of course, but it continued to serve the government because the tax leasing business, like the modern market in government bonds, was highly profitable.

In 1774 each member of the Company received a salary of 24,000 livres plus 4,200 for expenses. Interest on the 1,560,000 livres of capital required of each member was 10 percent for the first million and 6 percent for the remainder. In many cases much of this interest income had to be passed along to sleeping partners who provided capital to members of record. Much less is known about incomes realized from the distribution of profits. What can be stated with confidence is that the Company as a whole did not suffer a loss from 1726 to 1789. Yet it is also true that, while members enjoyed steadily growing profits, their incomes from the General Farms were not as fabulously high as their detractors imagined.

It has been estimated that between 1768 and 1786 Lavoisier received 1,200,000 livres in profits from the General Farms. Along with the other farmers, he benefited from a "long summer" of financial growth fueled primarily by the sustained expansion of the French economy. As economic activity increased, the income from indirect taxes on a variety of exchanges grew proportionately. The government was aware that had it been more effective in capturing these increased revenues, the Company would have profited less. By 1780 efforts were being made to bring the collection of indirect taxes under greater ministerial control. But by then royal finances were in such dire straits and the king's dependence on the financiers was so great that administrative reform alone proved incapable of breaking the financial deadlock.

Since it was widely believed that the financiers were a small group of private citizens who were holding the government hostage while gobbling up taxes, it is hardly surprising that they were targets of envy, resentment, and calumny. As tax collection became more efficient, those obliged to pay more knew just who to blame, and, as the wealth of some of the financiers began to outstrip that of the older aristocracy, the traditional structure of French society was placed under increasing strain. In 1774 a defender of the old order mingled envy with contempt when condemning the selling of positions as the king's secretary to rich financiers.

> Let a financier amass a lot of money, let him pile up riches on riches, that's his trade . . . but his fortune should be his only reward . . . It is odious that a man of nothing, often still disgusting and dripping with the mud from which he has emerged, may be a source of nobility by this office.[8]

Lavoisier could not have been unaware of such resentment and he clearly tried to conduct himself in public so as not to ignite the violent passions stirred up by the success of the financiers.

The relations between fiscal and political power in France between 1726 and 1789 were not typical of the nation's longer historical experience. The Company of General Farmers' prolonged monopoly, its control over all indirect taxes, its emphasis on vigorous and efficient collection, and its relative immunity from political intervention were all anomalous. When earlier French kings believed they were being ill-served by their financiers, they established special Chambers of Justice that restored the proper balance by reasserting royal authority. Debts were often repudiated and financiers hanged. No such courts were set up under Louis XV or Louis XVI. Hence it appeared that the king, who was the political representative of all Frenchmen, had been rendered powerless by the financiers. The laws of economics and contracts, and the sanctity of private property, offered little protection to the possessors of great wealth when they were later charged with plundering the nation's wealth and thwarting its will.

The French Revolution was born in a moment of political disarray that led to the fiscal collapse of the monarchy. This bankruptcy was the result of political inaction rather than greed or administrative incompetence on the part of the king's ministers and their creditors. The royal government was unable to resolve its financial problems because it had failed to live within its income, because it had failed to rationalize the allocation of tax burdens, and because it had failed to capture its full share of the growing revenues provided by existing taxes. The revolution that followed swept away the institutions of the old regime; the Royal General Farms passed from the scene forever in 1791.

The financial problems that had plagued royal government could not simply be eliminated by pulling down the old order, however, and revolutionary governments found themselves repeatedly obliged to address the problem of state finances. With boundless confidence in their ability to legislate solutions, they thoroughly repoliticized the administration of public finance. But in the absence of political stability, this way of dealing with the problem proved to be ineffective. As the financial needs of the revolution increased, the search for scapegoats and the use of political power to command financial resources intensified. Thus it came as no surprise, dismaying as the outcome was, that in 1794 Lavoisier, his father-in-law, and thirty other members of the Company of General Farmers were accused of having conspired against the nation. The revolutionaries were resurrecting in substance if not in precise legal form the earlier royal Chambers of Justice. In finding members of the Company guilty of conspiracy, the revolutionary tribunal revealed how partial and ultimately unsuccessful were eighteenth-century attempts to separate finances and politics.

Lavoisier joined the Company of General Farmers in May 1768 by buying a one-third share from an elderly member, François Baudon.[9] He was soon assigned several duties. His primary responsibility was to inspect the tobacco factories, warehouses, and retail shops in the provinces to the east and northwest of Paris and report on the performance of the customs agents charged with enforcing the tobacco regulations. He made his first tour in the autumn of 1768 and forwarded his reports to his future father-in-law Jacques Paulze, the head of the Tobacco Committee.

In May 1771 Lavoisier increased his share of Baudon's membership to one-half, and when the Alaterre lease expired in 1774, he signed the new David lease as a one-half member with Baudon. Shortly thereafter Turgot became the Controller-General of Finances and Lavoisier became much more active in the internal administration of the General Farms. When Baudon died in 1779 Lavoisier assumed his entire membership, although Baudon's widow continued to receive one-third of the profits for the remainder of the lease. Lavoisier and Baudon's widow agreed that she would be paid according to the profits realized in the preceding years of the lease, yet when the accounts of the David lease were finally settled in 1787, he voluntarily gave her an additional one-third of the unanticipated profits realized.

The letters Lavoisier wrote to Paulze between the latter part of 1768 and the early months of 1771 offer a number of insights into his character and conduct during these formative years. Lavoisier attacked the problems of the Farms' tobacco enterprise with vigor, optimism, and intelligence. He welcomed Paulze's advice, which he said "effected a revolution" in his understanding of the problems he faced, and he believed that, if the agents who collected sales taxes could be persuaded to supervise the tobacco retailers, it would bring about "a happy revolution" in the administration of the tobacco concession.[10] He was bothered that so many agents were intimidated by their superiors and were motivated by fear alone, and he reported that he was trying to cultivate a sense of competition and a concern for the overall good of the Company. When encouraging agents in different branches of the General Farms to cooperate, he tried to get them to see that reconciliation was in their own best interests, and when reorganizing administrative arrangements, he wrote out full justifications of his proposals because "one deals from strength when one has men of reason on one's side."[11] He realized, of course, that persuasion would not always succeed and that recalcitrant individuals had to be compelled to acknowledge the authority of their superiors and follow orders. Lavoisier said he had no patience with those who opposed what was good for the Company by insisting on what they claimed were their traditional rights.[12]

Lavoisier was aggressive in detecting and suppressing the sale of adulterated tobacco, but he was troubled that such tobacco often sold well. By adding ashes to milled tobacco, many retailers indulged in a kind of cheating that was "as injurious to the public as it is to management."[13] Lavoisier considered this the most serious administrative problem he faced: "I tell you the Company has good reason to think the retailers are real enemies of the tobacco concession. I am disturbed that circumstances compel me to report repeated evidence of this."[14] Chemistry, however, provided one way of dealing with this problem.

> Happily chemistry provides a reliable and unequivocal test that enables one to detect ashes in substances with which they have been mixed. If one moistens ashes with spirit of vitriol, aqua fortis, or any other acidic solution, a strong effervescence accompanied by a noticeable sound immediately appears. Since ashes mixed with tobacco do not lose this property, this test provides a way of determining if the tobacco has been adulterated with ashes. The greater or lesser strength of the effervescence

also provides an indication of the degree to which the tobacco has been adulterated with ashes.[15]

Armed with this test Lavoisier marched through his territory, examining the tobacco stocks of his retailers with the same diligence he devoted to the examination of well waters when touring with Guettard. The image is arresting: the all-powerful inspector from the capital, physically slight and formal in appearance, brisk and purposeful in manner, and accompanied by armed guards, enters a village shop and asks for samples of the tobacco sold to the public. He takes from his bag a vial containing a mysterious liquid which he pours on each of the samples. If any of them bubbles and foams, he declares it adulterated, a charge the merchant knows full well may result in the closing of the shop and the termination of his or her livelihood. But what if the merchant is innocent? Suppose the tobacco had been adulterated by those who supplied it to the shop? And why should the merchant be punished if his or her customers have not complained? But the inspector insists the test is infallible. The wretched merchant, whether guilty or innocent, has no recourse against the combined authority of science and the Company.

While Lavoisier was implacable in his pursuit of adulterated tobacco, he also knew the company could not flourish without retailers and the goodwill of its customers. He therefore showed a pragmatic compassion and restraint when disciplining those caught selling adulterated tobacco. Consider, for instance, the account of how he handled the difficult case of the widow Ducloir that Lavoisier forwarded to Paulze.

The tobacco sold by the widow Ducloir was not heavily adulterated, but it did produce a considerable effervescence when moistened with spirit of vitriol. And yet, while this shopkeeper's deception was adequately demonstrated, I would like to lay before you certain considerations you might bear in mind when deciding whether to dismiss her. 1. She is a widow of a former Captain General [of the General Farms] and she was given her commission to sell tobacco as a compensation. 2. She is responsible for a large family and has no other source of income. 3. She is especially close to Mr d'Auteroche. 4. She sells a great deal of tobacco. These consideration seemed to me sufficient to justify my sending for the widow Ducloir before leaving Chalons. I informed her the Company had been told she had adulterated her tobacco and that I was instructed to tell her she was dismissed. Frightened by this news, she told me her permit to sell tobacco was a reward for her husband's services and that if it were revoked she and her family would be reduced to poverty. After repeatedly turning aside her pleas and telling her how difficult it was to excuse such a serious infraction, and after she had repeatedly promised that we would never again have reason to complain about her accounts, I agreed to write in her behalf to the Company. It is now up to you to decide if the widow Ducloir was adequately punished by the fright I gave her. Her fate is in your hands.[16]

Lavoisier was also determined to rationalize the internal supervision of tobacco dealers. His "first principles" in this regard, he told Paulze, involved establishing a comprehensive and exact accounting system that would enable officials of the Company to trace the movement of each parcel of tobacco from the time it was acquired until it was delivered to the customer.[17] Careful weighing and extensive record-keeping were essential if the quantity and quality of the tobacco were to be protected. He realized that those in charge of the General Farms' tobacco warehouses, mills, and shops would not welcome the additional work required of them or closer scrutiny of their operations, but putting such a system in place offered the greatest possible security for both retailers and managers.

Administration and science for Lavoisier were clearly two sides of a common coin of rational procedure. As Charles Gillispie has observed, "to finance, munitions, and science alike he brought the luminous accuracy of mind that was his signet, the spirit of accountancy raised to genius."[18] When pursuing his cartographic project with Guettard, Lavoisier used the instrumental approach of physics and chemistry in his scientific field work. When on tour for the Company of General Farmers, he mastered the skills of accounting, negotiation, and personnel management. He made use of his scientific skill and knowledge whenever possible when addressing managerial problems, just as he used his managerial skills and knowledge when seeking to advance his vision of science. His handling of the case of the widow Ducloir gave him experience that proved valuable when he was called upon to report on scientific and technical claims forwarded to the Academy for appraisal. His accounting system for tobacco mirrored the system he used to account for the properties and weights of ingredients in chemical reactions. While seeking to persuade middle-level officials in the General Farms to cooperate, he acquired negotiating skills that he deployed to great advantage when serving as an administrator in the Academy of Sciences, as a director of the Gunpowder Administration, as a leading organizer of the Commission on Weights and Measures, and as a member of innumerable ad hoc committees. Lavoisier was more than a scientist who also took on administrative work or an administrator with an active interest in science. He was, more fundamentally, an unusually competent, ambitious, energetic, and methodical man whose personal style was marked by a high level of public spirit and a great respect for detail and rational procedure.

Lavoisier's program for reform, like so much of French culture, was radically disrupted by the revolution that began in 1789. It is hardly surprising, therefore, that those living in the post-revolutionary epoch have found it difficult to accept that some prominent figures in the old regime were pursuing essentially modern programs of reform. If Lavoisier's strategy for administrative reform incorporated so many principles we still consider valid, why did it fail? And how in a pre-democratic society could a man so privileged in his personal circumstances and so successful in his pursuit of wealth and fame have been sincerely devoted to the public good? We would find it simpler were we able to treat him as a clever habitué of Parisian salons who hawked visionary utopias of the sort concocted by so many enlightened *philosophes*. We would be more comfortable were we able to distance ourselves from Lavoisier, as post-revolutionary social theorists have distanced themselves from the Enlightenment, by portraying him as a well-intentioned but ineffectual individualist who was incapable of transcending the ideology of a self-serving elite. To do so, however, we would have to ignore most of the available historical evidence.

Although the historical record, being incapable of speaking for itself, must always be interpreted, it is not infinitely malleable. We are obliged by what is known about Lavoisier to acknowledge that he was more than a philosophical reformer. Lavoisier was not unaware of nor uninvolved with the world around him. While inspecting the operations of the General Farms, he worked hard at rationalizing its procedures and improving its internal discipline, yet he also knew that when other members of the Company were on tour they occasionally indulged in such demoralizing behavior as seducing employees' wives.[19] He also knew how to distinguish between genuine and spurious appeals to the common good. His brief profiles of officials he encountered on his tours are vivid and revealing. As he wrote of one of them,

> I have heard him repeatedly invoke the interests of the province, but after examining his ideas and reflecting on the conversion of taxes he has proposed, I have concluded that what he calls the public interest is nothing more than his own and that almost without exception the kind of man who proclaims himself the people's protector is attempting to do nothing more than transfer his own tax burden to the citizens of the province.[20]

Lavoisier was no Dr Pangloss. When he proposed the conversion of an archaic duty into a more convenient cash payment or the elimination of an odious special tax on Jews, he was motivated by considerations of both justice and administrative efficiency.[21] Perhaps, given his circumstances, he found it easy to be incorruptible. What is of greater moment was his conviction that reason mattered in public affairs, but only if coupled to personal virtue and attention to detail.

Lavoisier sought to improve the world that existed rather than imagine a perfect world, and he worked at his self-imposed task unstintingly. He did not overestimate the forces of order and reason, although they were his most formidable weapons. He was fully aware of the need to establish shared goals, to address questions of motivation, and to encourage persuasion and cooperation. One can, of course, invoke the logic of history and insist that the success of the Revolution and the failure of Lavoisier's efforts demonstrate the historical inadequacy of his program of reform. What cannot be questioned, however, is the sincerity of his commitment to reason and the public good or the vigor and intelligence with which he sought to improve public administration in France.

In championing the public good and the administrator's role in achieving it, Lavoisier believed he was simply carrying forward the program of reform articulated by a predecessor he considered one of the greatest figures in his nation's history, the seventeenth-century minister Colbert. Lavoisier expressed his appreciation of Colbert in several drafts of an *éloge* he intended to submit to the Académie Française for a competition to be judged in 1772.[22] He may well have hoped that this *éloge* would prepare the ground for his entry into the company of the forty immortals, just as his essay on street lighting had prepared his way for election to the Academy of Sciences. What greater honor could Lavoisier have aspired to than to join d'Alembert and Buffon in the most ancient and prestigious of all the royal academies? He did not complete his essay on Colbert, however, and the notes and drafts that survive must not be taken as representing his final thoughts on the subject. Nonetheless, what he says is revealing.

> Colbert saw the issue of taxation in a new way. He saw that most of the king's revenue consists of a partial interest in the public wealth, that the king, through taxation, has a greater or lesser interest in every kind of commerce without being directly responsible for it, from which he concluded that everything that tends to increase commerce at the same time contributes to the royal finances.
>
> In the midst of the chaos that surrounded him and sustained by his own courage and the profound truth of his perceptions, he conjoined the law to the good, without, like his predecessors, being burdened by outmoded relationships. He reasoned that while such relationships may be appropriate for determining questions of personal fortune and the life and honor of citizens, it would be dangerous to subordinate the principles of political administration to them. He was not concerned whether this or that tax pertained to the king's domains or whether it was old or new; what mattered was whether it was a burden to the people, whether it hindered the collection of taxes on other forms of revenue that were more abundant or more easily taxed.

What Lavoisier proposed, and what he believed Colbert had also attempted to achieve, was a functional rationalization of royal finances within a society that preserved its traditional hierarchies. It was a radical program in the seventeenth century and it remained radical in the eighteenth, even though when viewed from a post-revolutionary vantage point it looks socially conservative. In advocating Colbert's program for administrative rationalization of the state's finances, Lavoisier was clearly setting himself in opposition to those members of the Church and aristocracy, including the aristocracy of the robe, who regarded their tax exemptions as sacred and indefeasible rights.

Although an administrator rather than a politician, Lavoisier believed the nation could be reformed only under a king who enjoyed undisputed political authority. Lavoisier, like Colbert, was prepared to work for and live with the degree of centralization required to separate issues of taxation from issues of social status and political rights. In the event, however, the program proved to be politically unworkable; the king and his ministers could not carry it off. It is not hard to understand, however, why a young man of Lavoisier's talents and temperament found this program of reform appealing at a time when the political authority of the Bourbon monarchy appeared unshakable.

.

Part II

Consolidation and Contestation: 1775 to 1789

6

A New Theory of Combustion

Early in the summer of 1783, shortly before his fortieth birthday, Lavoisier and his brilliant collaborator, Pierre Simon Laplace, presented two memoirs to the Academy. The first of these was their jointly authored "Memoir on heat," which Laplace read at two meetings during the latter half of June. This major synthetic essay summarized the results of several years of research on the role of heat in the formation of vapors and the fixation of air. It opens with a detailed description of the new instruments and techniques Lavoisier and Laplace used to quantify the flow of heat in chemical reactions. It then describes specific experiments they performed and the conclusions they reached, and it ends with a wide-ranging discussion of the implications of their findings for theories of combustion and respiration. Firmly based on experimental data and profound in its theoretical significance, the "Memoir on heat" succeeded, as it was meant to, in establishing Lavoisier's command over a major area of research in eighteenth-century physical chemistry.[1]

At a meeting held between the two sessions devoted to the "Memoir on heat," Lavoisier announced the stunning news that, as he and Laplace had recently demonstrated to several members of the Academy, inflammable air (hydrogen) and vital air (oxygen), when burned together, form water. This experiment, he declared, established that water is not an element, but is, rather, a combination of two gases.[2] This discovery, along with the newly quantified theory of heat, appeared to provide irrefutable confirmation of Lavoisier's general theory of combustion. A decade after discovering that air combines with metals during calcination, Lavoisier had completed the construction of a comprehensive, coherent, and revolutionary new set of chemical theories.

There was another announcement during the summer of 1783 that also had interesting implications for the theoretical work Lavoisier was pursuing. Early in June in the town of Annonay the Montgolfier brothers conducted the first public launching of a hot-air balloon.[3] The government reacted to this wondrous event with an alacrity not normally encountered in bureaucracies, and on 2 July, a week before the invention of aviation was publicly announced in Paris, the Academy appointed a commission, with Lavoisier as a member, to investigate the Montgolfiers' aerostatic machine. At precisely that time Lavoisier was, for theoretical reasons of his own, trying to devise an experiment that would analyze water into its constituent gases, one of which is inflammable air. Like others familiar with the properties of inflammable air, the air given off when iron reacts with vitriolic acid, Lavoisier soon came to favor the use of this extremely light gas rather than hot air as the lifting agent for balloons. He therefore immediately realized that if the analysis of water could be performed cheaply and on a large scale, the invention of ballooning might prove to be of considerable public importance. Once again Lavoisier had the good fortune to be at the center of things when discoveries that raised interesting theoretical and practical possibilities were announced, and once again he seized the opportunities presented to him.

The chemical research Lavoisier pursued between the beginning of 1773 and June 1783 can be divided into two distinct phases. The second phase, during which he and Laplace put his theory of heat on a firm experimental foundation, started in 1777, when they began collaborating on the study of vaporization, and ended with the presentation of their "Memoir on heat." After completing this final phase of theory construction, Lavoisier was ready to campaign openly and aggressively for acceptance of his new chemical theories. Shortly after the "Memoir on heat" was presented he began to draft his "Reflections on phlogiston," a memoir destined to become one of the classics in the history of chemistry.[4] He read this rhetorical masterpiece to the Academy in 1785 and followed it up with a sustained, coordinated, and ultimately successful assault on the old chemistry. This campaign for the new chemistry will be described in greater detail in the next chapter.

Lavoisier began the first phase of the research he conducted between 1773 and 1783 with an intensive review of the published literature on fixed air; he brought this phase to a conclusion in 1777, when he presented his general theory of combustion to the Academy. He continued to pursue his underlying interest in the relations between physical changes of state and changes in chemical properties by concentrating on the fixation and release of air, and his study of fixed air thereby connected his methodological commitment to experimental physics and his desire to transform chemistry into an exact science. When presenting his 1785 "Reflections on phlogiston," the memoir in which Lavoisier launched his frontal assault on the received theories of chemistry, he said it was a sequel to the "Memoir on the general nature of combustion" (*Mémoire sur la combustion en général*) that he had read at the fall public meeting of the Academy eight years earlier.[5]

Between 1773 and 1777 Lavoisier pressed forward on many fronts. His foremost challenge was to integrate the properties of the newly discovered air that came to be called oxygen with what he had previously learned about the fixation of air. The story of how he met this challenge is especially illuminating, for it casts light on both the guiding assumptions he employed when studying natural phenomena and on the differences between his view of science and that of his great British contemporary, Joseph Priestley. To appreciate the full significance of this story, however, one must keep in mind that while oxygen is a natural substance that awaited discovery, the theoretical meaning of its properties was not self-evident. Lavoisier, unlike his contemporaries, quickly recognized that oxygen was destined to play a central explanatory role in the understanding of the phenomena he was examining. It took him many years and great effort, however, to turn this insight into a coherent and well-grounded set of theories.[6]

As we saw in chapter 4, Lavoisier presented his novel account of metallic calcination at the Easter meeting of the Academy in 1773. By the time he read this memoir he was deeply immersed in reviewing the literature on fixed air, an exercise that made him aware of just how complex a subject he had chosen to investigate. He was proceeding on the assumption that the base of air is, in chemical terms, a simple substance. He therefore expected solids and liquids that fix and release air to exhibit a consistent set of chemical properties, but nature was not cooperating. He noted with dismay, for instance, that calxes produced from calcareous substances, such as the quicklime obtained by roasting limestone, are formed by having their fixed air driven off, while metallic calxes are formed when metals combine with air. The fundamental properties of fixed air continued to elude him.[7]

He did not despair, however, as he continued to examine a wide variety of instances in which air is fixed in or released from solids and liquids. When his colleague Cadet reported to the Academy on 3 September 1774 that he had found a way to reduce the calx called the red precipitate of mercury to the metallic state without using charcoal, Lavoisier joined the committee that verified the experiment.* Unlike Joseph Priestley, who was at just this time determining that the air given off in this reduction exhibits some highly unusual properties, Lavoisier and his fellow academicians did not notice the special qualities of this air when confirming Cadet's observation.[8]

The next few months saw the unfolding of one of the most charming stories in the history of science.[9] Priestley visited Paris in November 1774. While dining with Lavoisier and other members of the Academy, he mentioned that he had found that the air separated from the red precipitate of mercury has some unexpected properties. Lavoisier had recently completed a detailed examination of Priestley's early experi-

---

* When metallic mercury is moderately heated in an enclosed vessel in the presence of air, a red precipitate (HgO) is formed. When this precipitate is heated intensely, oxygen is given off. When formed in this way the precipitate was called "mercurius calcinatus per se"; when formed from a compound of mercury and nitrate, it was called the red precipitate of mercury. Lavoisier gives a detailed description of his experiment in chapter 3 of his *Elementary Treatise on Chemistry*.

ments on different kinds of air and had commended them in print as

> the most formidable and interesting experiments on the fixation and release of air to appear since the work of Hales. I believe no other modern work shows so clearly the extent to which physics and chemistry offer new avenues of investigation.[10]

He therefore had good reason to take note of Priestley's informal report, and early in 1775 he again examined the air obtained from the calx of mercury. He collected a sample and tested it in several standard ways. He determined that it was not fixed air (carbon dioxide) and that it had most of the properties of common atmospheric air, but he also found that it is unusual in that it supports respiration and burning much better than common air. Lavoisier therefore concluded that it was elemental air in a highly purified form, "more pure than even the air in which we live."[11]

Lavoisier was delighted by the discovery of this purer form of air, for it seemed to provide a key piece in the puzzle he was contemplating. In 1773 he had discovered that air is fixed in metals during calcination; in 1775 he was prepared to provide a more precise account of that air. He did so in another Easter memoir titled "On the nature of the principle which combines with metals during calcination and increases their weight." This memoir has an interesting publication history. It first appeared in the May 1775 issue of Rozier's journal, and when Priestley saw it in that version he pointed out in print that Lavoisier had not fully appreciated just how novel the properties of this new air were. Lavoisier subsequently revised the memoir and read it again to the Academy in April 1778, and it was this revised version that was published in 1778 in the *Mémoires* of the Academy.[12]

Let us now take a closer look at Lavoisier's first version of this memoir and note in particular two passages he revised after reading Priestley's critique. In the first of these he distinguished forcefully between chemically pure air and the jumble of vapors contained in the atmosphere. The air that combines with metals, Lavoisier wrote, is

> neither one of the constituent parts of the air, nor a particular acid distributed in the atmosphere; it is the air itself entire and without alteration, without decomposition, even to the point that if one sets it free

after it has been so combined, it comes out more pure, more respirable, if this expression may be permitted, than the air of the atmosphere, and is more suitable to support ignition and combustion.[13]

When thinking about the properties of air Lavoisier was evidently making a categorical distinction between elemental air, which combines with metals to form calxes, and the mixture of vapors present in the atmosphere. To isolate and characterize elemental air, he studied reactions that extract that air from the chaos of the atmosphere; fixation is in this sense a natural process that isolates pure elemental air. As he wrote in a memoir on the calcination of lead that he read to the Academy in November 1774, "these experiments provide a means to analyze the fluid which comprises our atmosphere and to examine the principles which constitute it."[14] Lavoisier's goal was to construct explanations of chemical phenomena based on the enduring properties of highly reactive simple substances such as air, water, and fire; he was not attempting, as Priestley was, to catalogue the properties of the complex compounds and mixtures encountered directly in nature. His research program was more focused and, in a theoretical sense, more ambitious than Priestley's, and therefore he often ignored curious facts of no immediate theoretical significance. While trying to make sense of oxygen Lavoisier ignored the complexities raised by the presence of many different airs and vapors in the atmosphere. It was Priestley, not Lavoisier, who discovered that oxygen exists as a major component of the atmosphere; it was Lavoisier who determined the theoretical significance of oxygen for the science of chemistry.

Another passage that Lavoisier included in the 1775 version of his paper but completely recast three years later took issue with Priestley's use of the phlogiston theory to explain changes in the properties of airs.

> From the fact that common air changes to fixed air when combined with charcoal it would seem natural to conclude that fixed air is nothing but a combination of common air and phlogiston. This is Mr. Priestley's opinion and it must be admitted that it is not without probability; however, when one looks into the facts in detail, contradictions arise so frequently I feel it necessary to ask natural philosophers and chemists still to suspend judgment; I hope to be soon in a position to communicate the reasons for my doubts.[15]

This passage indicates that Lavoisier believed there was no necessary connection between the discovery of a purer form of air and the plausibility of the phlogiston theory. He had already determined that the phlogiston theory provided an account of the calcination of metals that was in some ways a mirror image of the truth. He was confident that in time he would be able to provide a fully articulated theory of the elements and their enduring chemical properties, a theory capable of replacing the phlogiston theory that Priestley and other chemists continued to employ so uncritically. By the spring of 1775 Lavoisier had reason to believe the campaign against the phlogiston theory was well advanced and that the older theory would be discarded as soon as well-grounded new theories had been developed to replace it. New discoveries and careful quantification would eliminate the phlogiston theory and transform chemistry into a true science. But we must be careful not to overrate Lavoisier's daring or originality in holding these views, for he was not alone in thinking that the demise of phlogiston was just a matter of time.[16]

The highly purposive Lavoisier clearly had little sympathy with Priestley's apparently aimless ransacking of nature, even though he respected this curious investigator's manipulative skill and keen perception. Lavoisier characterized Priestley's investigations, which he realized deserved careful scrutiny, as "a fabric woven of experiments that is hardly interrupted by any reasoning."[17] Having met Priestley, Lavoisier probably recognized him as one of God's fools, an investigator whose chief delight was to glory in the unfathomable wonder of creation.[18] Priestley, after commenting on several points in Lavoisier's 1775 paper and saying that he looked forward to the publication of his alternative to the phlogiston theory, airily devalued the whole enterprise in a manner that Lavoisier may have found either comic or scandalous.

> It is pleasant when we can be equally amused with our own mistakes, and those of others. I have voluntarily given others many opportunities of amusing themselves with mine, when it was entirely in my power to have concealed them. But I was determined to shew how little *mystery* there really is in the business of experimental philosophy, and with how little *sagacity*, or even *design*, discoveries (which some persons are pleased to consider as great and wonderful things) have been made.[19]

And yet Lavoisier was wise enough not to be put off by manner alone – perhaps he knew that the great astronomer Kepler had also been one of God's fools. Truths can be discovered in the most unexpected ways and Lavoisier shrewdly paid careful attention to all the observations reported by his prolific, chatty, and unsystematic British contemporary.

Although Lavoisier's vision was not blocked by Priestley's use of the phlogiston theory, he had considerable difficulty assimilating Priestley's discovery that approximately one-fifth of the atmosphere consists of this "dephlogisticated air" (oxygen). Priestley's claims about the composition of the atmosphere did not pose a direct challenge to Lavoisier's ideas, for he had no particular preconceptions on this issue. What disturbed him was that Priestley's discovery shifted the focus of investigation from the study of "eminently respirable air" produced in the laboratory to questions about how this gas can exist in a uniformly mixed state in the atmosphere. This shift threatened to disrupt the research program Lavoisier wished to pursue. So long as he could confine his research to chemical reactions that exhibit detectable weight changes, Lavoisier felt confident he could achieve the quantitative exactness needed to validate his theoretical claims. But if he was also obliged to provide a physical theory of the heterogeneous atmosphere, he was much less sure he could succeed.

Priestley forced this issue in a somewhat confusing set of comments on Lavoisier's 1775 paper that were published in the last section of the second volume of his *Experiments and Observations on Different Kinds of Air*. "Mr. Lavoisier," he wrote,

> as well as Sig. Landriani, Sig. F. Fontana, and indeed all other writers except myself, seems to consider common air (divested of the effluvia that float in it) as a simple *elementary body*; whereas I have, for a long time, considered it as a *compound*; and this notion has been of great service to me in my inquiries.[20]

To understand the meaning of this statement and the challenge it posed to Lavoisier's research program, we must step back a few years to examine the nitrous air test for the "goodness" of common air that Priestley had described at length in the first volume of his *Experiments and Observations*.[21] This experimental test, which quantified certain

140

changes of state volumetrically rather than gravimetrically, focused Priestley's attention on the composition of the atmosphere. When this test became linked via oxygen to the study of the fixation and release of air, it forced Lavoisier to extend his research program in ways that it seemed might prevent him from bringing it to a successful conclusion.

Priestley turned to the study of airs fairly late in his career and when he did so he built upon the earlier research of Stephen Hales.[22] Hales had noticed that when the air given off by the reaction between metals and aqua fortis is mixed with common air, a "red fume" forms and the volume of the gases decreases.* Priestley was looking for new species of air and hence examined this reaction closely. He began by measuring the extent to which the volume of "nitrous air," as he called nitric oxide, was diminished when mixed with other gases. When mixed with fixed air (carbon dioxide), inflammable air (hydrogen), or "diminished air" (mostly nitrogen), the nitrous air experienced no reduction in volume. Priestley also determined that the gases that remain after atmospheric or common air has been mixed with nitrous air will not support respiration. He concluded, therefore, that the reaction "is peculiar to common air, or air *fit for respiration*."[23] Further experiments indicated that the degree of diminution ranged from zero for airs which would not support any respiration to "one third of the whole" for those that best supported respiration. "We are, by this means," Priestley concluded in the first volume of *Experiments and Observations*, "in possession of a prodigiously large *scale*, by which we may distinguish very small degrees of difference in the goodness of air."[24]

Instrumentally, Priestley's test for the goodness of air suggested the construction of a piece of experimental apparatus that came to be called the eudiometer, a device used to measure the changes in volume that occur when gases combine. The earliest such instruments were developed by entrepreneurial chemists who used Priestley's test for the goodness of air to identify locations having salubrious climates. But the eudiometer was also used by disinterested scientists who wished to study what happens when gases combine. Priestley and Volta, among

---

* In modern terminology, nitric oxide, the vapor given off when metals are dissolved in nitric acid, combines with atmospheric oxygen to form nitrogen dioxide, which is reddish in color and highly soluble in water.

others, employed modified forms of this instrument to examine the effects of detonating various gases by an electric spark and of burning together such gases as eminently respirable air and inflammable air. And as we shall see below, it was experiments made with the eudiometer that led directly to Lavoisier's dramatic announcement that water, rather than being an element, is a compound of oxygen and hydrogen.[25]

Priestley's test for the goodness of air, although quantitative, could not be easily assimilated into Lavoisier's gravimetric research program. When reviewing the first volume of Priestley's *Experiments and Observations*, Lavoisier devoted an entire section to the nitrous air test.[26] He recognized the test's importance and used it early in 1775 when examining the air driven off from the red precipitate of mercury. And yet the test puzzled him, for it appeared that the dramatic reduction in the volume of air was not matched by a comparable change in specific weight. As Lavoisier wrote in his 1774 *Physical and Chemical Essays*,

> It is a singular and almost unbelievable fact that nitrous air, whether by itself or combined with common air, always has a specific weight roughly equal to that of atmospheric air. In a volume of three pints Mr Priestley never found more than a half grain of difference, more or less. But how are we to understand a situation in which two fluids become so mingled that their volume is reduced by a third without the specific weight of the mixture being greater than that of either of the two fluids?[27]

Lavoisier, who a few years earlier had invested so much hope in examining the specific weights of salt solutions, did not know what to make of such a dramatic condensation of volume without a comparable increase in density. His puzzlement clearly indicates that he felt blocked by not being able to transform Priestley's volumetric analysis into a more precise gravimetric analysis.[28] The problem took on even greater importance when Priestley announced he had discovered that respirable air in its pure form exists as a separate gas in the atmosphere.

Priestley discovered the gas Lavoisier named oxygen while experimenting with the air given off by the red precipitate of mercury. He began by subjecting this air to his nitrous air test. When he examined the residual air, he found to his surprise that it would still support respiration. He therefore added more nitrous air to the sample. Whereas one volume of nitrous air produced a maximum diminution of atmospheric air when mixed with two volumes, it took more than four volumes of nitrous air to react completely with a single volume of the air given off by the red precipitate of mercury.[29] Priestley concluded that the air given off by the red precipitate of mercury is several times "purer" than the atmosphere. He also pointed out that since the atmosphere contains this air, it must be a mixture of at least two kinds of air. How two or more gases having different chemical properties and, presumably, different specific weights could form a homogeneous mixture was not a problem that interested Priestley, but Lavoisier did worry about it, as did John Dalton after him.[30]

Lavoisier found 1775 a difficult year. At the beginning of the year he looked forward to making a quick conceptual breakthrough in his struggle to explain the chemical phenomena associated with the fixation and release of air. This expectation was further encouraged by the discovery of a form of air that was purer than common air. But by the end of the year the research problems he was focusing on had become more complex rather than less. Chemists were isolating and describing more and more gases, yet he had not managed to formulate a coherent theory that could explain even the most prominent properties of the well-known airs; also, he had not articulated an account of the vapor state that was more exact and more firmly grounded on experiment than that derived from the phlogiston theory. He had hoped that soon all the pieces needed to construct a comprehensive replacement for the phlogiston theory would be in hand and that a new theory could be assembled with a minimum of effort, but that was not to be. The study of airs offered no easy victories.

Having been thwarted in his search for a comprehensive theory, Lavoisier concentrated on more restricted problems. Perhaps by developing more limited concepts, instruments, and techniques of investigation he could begin working his way up to a general theory. He did not intend to present to his colleagues theories that could not meet the exacting standards of proof expected of an experimental physicist. Priestley's use of the nitrous air test to discover "eminently respirable air" in the atmosphere forced Lavoisier to employ new modes of analysis and consider problems he had hoped to avoid. Yet he also seems to have understood quite early on that the identification of this new kind of air might enable him to turn some of his favorite hypotheses into theories. To do so he would have to perform many new experiments and devise persuasive arguments. Once he had become familiar with the chemical properties of oxygen, however, he quite quickly found ways to demonstrate that it plays a central role in the formation of acids, in respiration, and in combustion.

He began with the origin of acids. Lavoisier, like many of his contemporaries, believed that the cause of acidity in all its many forms could be traced back to the properties of a single substance. He knew that the calxes of sulfur and phosphorus form powerful acids when mixed with water and that these calxes contain air that was fixed in them during their formation. He therefore found it reasonable to assume that all acids contain a great deal of air, that air is the common substance in all acids, and that the differences between acids arise from the different specific principles they contain. He stated all this as background at the beginning of the "Memoir on the presence of air in nitrous acid" that he read to the Academy in April 1776.[31]

The task Lavoisier set himself in this memoir was to demonstrate that he could successfully apply his general hypothesis on the origins of acidity to the properties of nitrous acid. This acid was of special interest both because it played a central role in the manufacture of gunpowder, a subject to which Lavoisier had recently begun to devote a great deal of time, and because of the ways in which Joseph Priestley had used it. In early 1776 Lavoisier was still struggling with the implications of the nitrous air test. Having no desire to provoke a controversy, he began with a disclaimer:

> Before taking up my proper subject, I wish to state that I was not the first to perform several of the experiments presented in this memoir; indeed, strictly speaking, only Mr Priestley can claim originality in this regard. However, since the same facts have led us to diametrically opposed conclusions, I trust that if I am criticized for having borrowed demonstrations from this celebrated scientist, the originality of my conclusions will not be challenged.[32]

For our purposes the most important of Lavoisier's conclusions is his claim that pure air is the source of acidity in nitrous acid. By analyzing the acid he had, he wrote, "demonstrated that it contains air, or rather an air that is pure and (if one may say so) more airy that common air."[33]

Clearly Lavoisier considered this pure air elemental air itself. "It appears to have been proven," he wrote, "that the air we breathe contains only one-quarter true air."[34] And it is this element, he concluded, that makes nitrous acid acidic. When the acid reacts with a metal, "the metallic substance takes hold of the portion of pure air contained in the nitrous acid that constitutes its acidity."[35] Lavoisier first used the term "oxygen," a neologism meaning the begetter of acids, in a synthetic memoir on the nature of acids he submitted to the Academy in September 1777.[36] As we have seen, however, a year and a half earlier he had publicly and with complete conviction asserted that the gas he later called oxygen is the cause of acidity.

And what of Priestley's views on these matters? Lavoisier ended his 1776 memoir with two arguments that demonstrated he too could detect and correct the errors of learned colleagues: the first addresses the phlogiston theory, the second Priestley's use of the seventeenth-century idea that respiration and combustion are supported by a nitrous substance in the atmosphere. Concerning the phlogiston theory, Lavoisier said:

> It will be asked whether the phlogiston of the metal plays a role in this operation [of collecting pure air by forming the red precipitate of mercury]. While not trying to settle a question of such importance, I would reply that when the mercury leaves the operation exactly as it entered it, there is no evidence that it has lost or gained phlogiston, at least if one does not suppose that the phlogiston involved in the reduction of the metal passed through the walls of the containers. To do that would be to suppose a special type of phlogiston, one quite different from that of Stahl and his disciples. That would be a return to the principle of fire, of fire combined with substances, a much older and very different view from than of Stahl.[37]

Lavoisier was equally polite but severe with Priestley's views on what it is that enables the atmosphere to support respiration and combustion.

> I will end this memoir, as I began it, by crediting Mr Priestley for most of whatever it contains that is of interest. Yet the love of truth and the progress of knowledge to which we are dedicated obliges me to point out an error he has committed, one that should not pass uncorrected. This celebrated physicist has noticed that when nitrous acid combines with an

earth, the latter takes its common air or its air that is better than common air. From this he has concluded that atmospheric air is a product of nitrous acid and earth. This hardy idea has been adequately disproven by the experiments presented in this memoir. It is clear that air is not composed of nitrous acid, as Mr Priestley supposes, but rather that the nitrous acid is composed of air. This single remark provides the key to a great many of the experiments described in sections III, IV and V of Mr Priestley's second volume.[38]

By November 1776 Lavoisier was prepared to criticize another of Priestley's recent theoretical accounts, the subject this time being the role of air in respiration. After drafting a sketch of his ideas, Lavoisier asked that his memoir on the decomposition of air in the lungs be listed for presentation at the Academy's fall public meeting. The schedule was already full, however, and he had to wait until the next Easter meeting to present his new theory of respiration. The memoir he read at that meeting, "Experiments on animal respiration and the changes air undergoes in passing through the lungs," marked another step in his growing mastery of the gases of the atmosphere, in his rejection of the phlogiston theory, and in the construction of his general theory of the role of oxygen in combustion.[39]

In January 1776 Priestley described to the Royal Society in London a series of experiments he had performed on the effects of different gases on the blood. Physiologists in the seventeenth century had asked why the blood regains its vivid arterial color in the lungs; Priestley sought to answer the question by investigating the effects of dephlogisticated air (oxygen) on blood. When he exposed samples of a dark blood clot to dephlogisticated air, he found that the pieces became a florid red, while the air to which they were exposed, when tested with nitrous air, was seen to have lost its ability to support respiration. This illustrates, Priestley concluded, that it is phlogiston that makes the blood black and that blood is revitalized in the lungs by getting rid of its phlogiston. Lavoisier, thanks to the intelligencer Magellan, received a copy of Priestley's printed report of these experiments in April, just as he was completing his work on the air in nitrous acid and turning his attention to the problem of respiration. Once again he found much to admire in Priestley's innovative experiments and much to criticize in his interpretations.

Lavoisier noted that Priestley "performed very ingenious, very delicate and highly innovative experiments by which he attempted to
prove that animal respiration, like the calcination of metals and other
chemical processes, phlogisticates the air and that the air ceases to be
respirable when it is charged and saturated with phlogiston."[40] But if
Priestley had identified an interesting problem and performed some
well-conceived and relevant experiments, as Lavoisier believed he had,
his theoretical explanation of the phenomena observed was hardly
credible.

> No matter how plausible the theory of this famous physicist may appear
> on first encounter and no matter how numerous and well performed are
> the experiments he offers to support it, I find it is contradicted by so
> many phenomena that I believe it is fatally flawed. I therefore adopted a
> different approach and have found myself unavoidably led by my ex-
> periments to conclusions utterly opposed to his. I will therefore not
> pause here to discuss in detail each of Mr Priestley's experiments or show
> how they can be seen to support the view I will advance in this memoir.
> I will only describe those that seem most germane and give an account of
> their results.[41]

Lavoisier continued to learn from Priestley, but he no longer felt any
obligation to respond to his advocacy of the phlogiston theory. Although the new regime of chemistry was still being constructed, the old
regime, at least as represented by the phlogiston theory, was in
Lavoisier's view no longer viable.

The experiments and arguments Lavoisier presented in his memoir
on respiration reveal just how far he had gone in bringing order to
pneumatic chemistry. He performed simple experiments analyzing atmospheric air into its salubrious and mephitic parts and then recom-
bined these airs. Such experiments, he asserted, "provide the most
complete type of proof that one can attain in chemistry, the decomposition of air and its recomposition."[42] He also distinguished between
the air rendered mephitic by respiration (carbon dioxide) and the non-
respirable portion of the atmosphere, which he called *mofette*. Since the
air given off in respiration is the same as the air driven off when lime
is roasted, it should be given a proper chemical name rather than being
called "fixed air." And since the base of this air, that is to say the part
that combines with fire in vaporization, forms a weak acid when mixed

with water, Lavoisier gave this gas the new name of chalky acid in the aeriform state (*acide crayeus aériform*).

Lavoisier was of two minds on the subject of respiration itself; respirable air was either converted into chalky acid in the lungs or it was carried throughout the body by the blood. The fact that the chalky acid can also be formed by burning charcoal appeared to support the view that respiration is a combustion in the lungs. He was inclined to accept the second view, however, because of a close analogy based on color. This rather surprising line of reasoning reconnected Lavoisier's theory and Priestley's experiments. It also constituted an attempt to bring within the domain of his emerging theory of oxygenation a set of chemical phenomena, that is to say color changes, that other chemists explained in terms of the presence or absence of phlogiston.

> It is known that one of the properties of eminently respirable air is to communicate the color red to substances, above all the metallic substances with which it is combined: examples are mercury, lead, and iron . . . The same effects . . . are found . . . in the calcination of metals and animal respiration.[43]

The blood, Lavoisier argued, picks up eminently respirable air and carries it to the rest of the body; that is why it becomes bright red in the lungs. This was not his last word on respiration, but even at this rudimentary level of development this theory enabled him to separate Priestley's empirical findings, which he respected, from his theoretical interpretation of them, which he rejected.[44] If Lavoisier was going to construct a general theory of combustion, he had to be able to provide a plausible account of the role of air in respiration, one that did not depend on the phlogiston theory. After 1776 he was prepared to do so.

On 12 November 1777 Lavoisier read to the Academy a brief synoptic memoir that summarized his theoretical views on air. In this "Memoir on the general nature of combustion" he restated his methodological principles, openly attacked the Stahlian phlogiston theory, and related the experiments he had performed and the theories he had been developing to the larger topic of combustion. After two and a half years of criticizing Priestley's advocacy of the phlogiston theory and promising to provide a more adequate theory of his own, Lavoisier was at last beginning to deliver the goods.[45] But as he acknowledged in this memoir, he had not yet performed all the experiments needed to turn his hypothesis into a theory. Despite these limitations, however, the "Memoir on the general nature of combustion" is a bold and forceful scientific statement.

Lavoisier began with a series of methodological assertions designed to put some distance between himself and Priestley on the one hand, and between himself and his more speculative French colleagues on the other.

> While the spirit of system is dangerous in the physical sciences, it is equally to be feared that the disorderly heaping up of a great many experiments will obscure science rather than clarify it, raise barriers to those who wish to advance beyond the first stages, and cause long and difficult research to lead to nothing but disorder and confusion. Facts, observations, and experiments are the building blocks of a great edifice, but when gathering them in science one must avoid creating obstacles. One should, rather, organize them and indicate the classes to which they belong and to which part of the whole each of them pertains.
>
> When considered from this point of view, systems in physics are merely instruments that serve to alleviate the weakness of our senses. They are, more precisely, methods of approximation we employ while solving problems. They are hypotheses which, after being altered, corrected, and changed whenever contradicted by experiments, will one day infallibly lead us, through exclusions and eliminations, to knowledge of the true laws of nature.
>
> Today, emboldened by these thoughts, I risk proposing to the Academy a new theory of combustion, or rather, to speak with the caution I should observe, a hypothesis that explains in a highly satisfactory manner all the phenomena of combustion, calcination, and even, in part, those that accompany the respiration of animals.[46]

Lavoisier concluded this prefatory statement by asserting that his singular hypothesis "directly contradicts Stahl's theory and those of many famous men who have followed him." Although not one to stir up controversy for its own sake, Lavoisier was eager to gather the harvest of his own efforts now that the time was ripe.

The new theory of combustion is grounded on four invariable facts that, according to Lavoisier, "appear to be laws that nature never violates." These are: (1) combustion is accompanied by the release of material fire or light, (2) combustion occurs only in "pure air," (3) during combustion pure air is decomposed and the weight of the burning substance increases in exact proportion to the amount of the air decomposed, and (4) during combustion the substance burned is changed into an acid by the addition of the substance that increases its weight. This theory attains great generality, Lavoisier explained, because of the analogy that exists between combustion, calcination, and respiration.

Lavoisier knew that those who had listened attentively to the various memoirs he had read to the Academy from 1773 to 1777 would find nothing surprising in this general theory of combustion. He could, of course, have offered his theory as a revision of received ideas that were relevant to only a limited set of problems, and he could have waited for it to be absorbed into the body of science with as little disturbance as possible. But from the outset Lavoisier had shown that he was too intellectually ambitious to settle for mere recognition as a prominent figure in a discipline that lacked integrity and coherence. He had bigger game in sight, and having stated his theory of combustion in general terms, he set out to demolish the phlogiston theory.

> The various phenomena associated with the calcination of metals and combustion can be adequately explained by Stahl's hypothesis, but to do so one must assume, as he does, that there is a matter of fire, a phlogiston fixed in metals, in sulfur and in all those substances considered combustible. If, however, one asks the proponents of Stahl's doctrine to prove that this matter exists in combustibles, they inevitably fall into a vicious circle. They must reply that combustible bodies contain material fire because they burn and that they burn because they contain material fire. Obviously in the final analysis this is to explain combustion by combustion.[47]

Was Lavoisier attacking the concept of phlogiston itself, or was he attacking the whole mode of explanation of which the phlogiston theory was but a notable example? The answer can be found by examining the alternative theory he proposed in 1777. It is, essentially, a transformed phlogiston theory, one in which the matter of fire is not located in the metal or in the combustible, but rather in the vaporous air that combines with them when they burn. In other words, Lavoisier's alternative theory, like the phlogiston theory, also relied on the properties of fixed fire, a substance that in 1777 he had not yet differentiated into heat and light nor begun to quantify.

> The existence of material fire, or phlogiston, in metals, in sulfur, etc. is really nothing but an hypothesis, a supposition, which, once it is granted, explains rather well some of the phenomena of calcination and combustion. However, if I can show that these same phenomena can be explained in a much more natural way by the opposite hypothesis, that is to say without supposing there is any material fire or phlogiston in combustible substances, then the Stahlian system will find itself shaken to its foundations.
>
> I will, of course, be asked what I mean by the matter of fire. I reply with Franklin, Boerhaave, and one group of philosophers of antiquity in saying that the matter of fire or light is a very subtle and very elastic fluid that surrounds all parts of the planet we live on, which penetrates with greater or lesser ease all the bodies of which it is composed, and which tends, when free, to distribute itself uniformly in everything.[48]

If Lavoisier's willingness to speculate in a public presentation to the Academy was a measure of his confidence in his research program, then in November 1777 he was feeling very confident indeed. Although he certainly could not prove that his aetherial theory of fire was true, he sought to render it plausible by appealing to a number of familiar analogies and definitions. He turned first to the analogy with salts.

> I will add, by borrowing from the language of chemistry, that this fluid [i.e. fire] dissolves a great many bodies, that it combines with them in the same way that water combines with salts and that acids combine with metals, and that bodies so combined and dissolved in fluid fire lose some of the properties they had before the combination and acquire new ones that make them more like the matter of fire.[49]

He also sketched out his organizing assumption that "all bodies can exist in three different states" and used it to support his notion that all substances in the vaporous state are compounds of the matter of fire and a specific base. He argued that, since the matter of fire is located in the air rather than in the combustible, we really ought to say that it is the air that burns: "Pure air, Priestley's dephlogisticated air, is, according to this view, the truly combustible body."[50] If one adopts this view, Lavoisier continued, the elasticity of gases can be easily explained in terms of the fixed fire they contain. Seeing combustion as the fixation of the base of pure air also eliminated the problem of having to explain how there can be a weight gain during combustion while phlogiston is being given off.[51]

Lavoisier argued with the confidence of a lawyer who has an airtight case, yet it is fair to ask what evidence he presented to demonstrate that fixed fire is set free during combustion. The evidence relevant to this question, he asserted, is to be found in the flames and light that accompany combustion. When the base of air combines with a substance to which it is intensely attracted, "the matter of fire recovers its rights, its properties, and reappears before our eyes with heat, flame and light."[52] But Lavoisier could not even convince himself that this kind of evidence was adequate, and so in the next to last paragraph of this extraordinarily speculative excursion, he insisted again that his alternative to the Stahlian theory was still only a hypothesis.

> I repeat that in attacking Stahl's doctrine I have not sought to offer an alternative theory that is rigorously demonstrated, but only a hypothesis which seems to me to be more plausible and more in accordance with the laws of nature. It appears to me to contain implicitly explanations that are less forced and contain fewer contradictions.[53]

If Lavoisier was willing to admit that not all his claims had been validated, his faith in what he knew to be true remained firm. He closed by promising to pursue further every facet of this theory and to report his findings to the Academy in future memoirs. But even before undertaking these investigations, he declared, "I dare to assert in advance that the hypothesis I have laid before you will explain in a satisfying and simple way the principal phenomena of physics and chemistry."[54] This was the voice of an investigator in full cry, one who is utterly convinced that his research is on the right track. Lavoisier was prepared to gather the harvest. Shortly after reading his "Memoir on the general nature of combustion" he began drafting the introduction and plan for a second volume of *Physical and Chemical Essays*.[55]

By the time he read the "Memoir on the nature of combustion," Lavoisier was already working on a more adequate theory of fixed fire or, as he came to term it, caloric.[56] His recent collaboration with Laplace on experiments concerning heat and vaporization had made him aware

152

of the younger man's abilities, yet while he was eager to pursue these topics with Laplace, Lavoisier was able to give only part of his time to the study of heat. In 1780–1 the situation changed, however, when French chemists suddenly became aware of the profound investigations into specific and latent heat that Joseph Black and his disciples had been conducting in Great Britain. Although nearly a decade earlier they had heard reports of Black's work, the publication in 1780 of a French version of Peter Dugud Leslie's *A Philosophical Inquiry into the Causes of Animal Heat* and the appearance of summaries of Adair Crawford's recently published *Experiments and Observations on Animal Heat* alerted them to the importance of the concepts and methods the British were utilizing. By midsummer 1781 a French summary of Crawford's book prepared by the ever-alert Magellan had been published in Rozier's journal. A few months later Lavoisier and Laplace conceived and designed the instrument, later called a calorimeter, they would use to provide a firm quantitative foundation for Lavoisier's theory of caloric.[57]

One can well imagine Lavoisier's state of mind when he first realized that certain prominent British chemists had stolen a march on him in the quantitative study of heat. Was he once again to be limited to responding to ingenious but confusing experiments and incoherent and distracting theories? How could he seize control of the production and interpretation of the experimental data that would enable him to complete the theory that would, he believed, explain the principal phenomena of physics and chemistry? Was another Priestley going to be visited on him, and if so, would he be able a second time to hold his research program together while responding to a deluge of disparate reports from abroad?

Lavoisier responded with vigor to these challenges. He had the means, the colleagues, and the will to press forward rapidly so as to make French studies of specific and latent heat the best in the world. He and Laplace designed new instruments and performed novel experiments, they measured the flow of heat in changes of state and in chemical reactions, and they produced a body of information that brought the quantification of heat exchanges to a high level of experimental precision.[58] When Laplace read their "Memoir on heat" to the Academy in June 1783, he was placing the final stone in the arch formed by Lavoisier's interrelated theories of acidification, calcination, combustion, and respiration. Lavoisier then adorned this arch with his dramatic announcement that water, long thought to be an element, is in fact a compound formed by the combination of two gases.

The story of how Lavoisier came to demonstrate that water is a compound of oxygen and inflammable air conforms perfectly to the well-established pattern of British discovery and French explanation.[59] According to Lavoisier's theory of combustion, a substance that combines with oxygen should form an acid, and in 1781–2 Lavoisier had in fact looked without success for the acid formed when inflammable air burns. Our next report of his investigation of this reaction comes from June 1783 when, with Laplace's assistance, he used a new piece of apparatus that had just been constructed to his specifications to determine what is produced when large quantities of hydrogen and oxygen are burned together.[60] This instrument consists of a large glass sphere fitted with a brass cap containing piping that permits the introduction of a steady flow of the two gases that are burning in the globe.[61] By using this elaborate device, along with the pneumatic chests that supplied it with pressurized gases, Lavoisier and Laplace were able to demonstrate, in an experiment whose results were quite approximate, that when these two gases combine they form pure water equal in weight to the weights of the gases consumed. That they found the result puzzling is made clear by a letter Laplace wrote four days later to the intelligencer Jean Deluc in London: "We obtained," he said, "pure water, or at least such as had no trace of acidity and was insipid to the taste."[62]

The experiment Lavoisier and Laplace performed on 24 June 1783 was witnessed by several French colleagues and a distinguished visitor from the Royal Society of London, Charles Blagden. Blagden had brought to Paris the news that his reclusive but experimentally talented English colleague, Henry Cavendish, had recently used an electric spark to combine inflammable air and dephlogisticated air in a eudiometer and that the water formed weighed exactly as much as the gases employed. The French scientists told Blagden they already knew Priestley had combined these two gases and found water. They expressed surprise, however, when told the water weighed as much as the total weight of the gases.[63]

This was a somewhat disingenuous reaction, however, for at least one report of Priestley's experiments that had reached Paris indicated the weight of the water formed was equal to the weight of the constituents. Of course Lavoisier and his colleagues may have doubted Priestley on this matter. In any case it was this claim they examined and confirmed with their new apparatus on 24 June. And it was this claim that Gaspard Monge, who had heard the reports of Priestley's discoveries while working with Lavoisier in Paris, also confirmed in a series of independent and precise experiments carried out at the same time in the military school of Mézières.[64]

The path to significant discovery is frequently marked with false leads and confusion. Late in 1782 Priestley had become convinced he had discovered a way to convert water into a new kind of air by combining it with quicklime and then exposing it to intense heating. Priestley wrote excitedly of his discovery to his friends and colleagues Josiah Wedgwood, James Watt, Edmund Burke, Arthur Young, and Sir Joseph Banks, telling them that the primordial atmosphere probably consisted of this air and that it became respirable only when enriched by the action of plants.[65] He reported in March that by pursuing a hint from Cavendish he had succeeded in turning air into water by igniting a spark in a mixture of inflammable and respirable air. In April 1783 he realized he had misunderstood his earlier experiment and that steam rather than a new air was the product. He described all this in a written communication to the Royal Society of London; by mid-May his report had been summarized and discussed in the Academy of Sciences in Paris.[66]

In the summer of 1783 Lavoisier, with the invaluable assistance of Laplace, brought the construction of his comprehensive new theory of combustion to closure. Whereas in 1777 he was still describing his theory as a hypothesis, in 1783 it was, so far as he was concerned, a well-established and completely credible theory. This is not to say that the fecundity of his research program had been exhausted, for, as Lavoisier realized, many interesting problems still awaited thorough investigation.[67]

Lavoisier's standards of demonstration obliged him to provide experimental confirmation for certain claims that by the summer of 1783 had not been fully verified. He had, for instance, demonstrated that oxygen and inflammable air combine to form water, but he had not yet analyzed water into these two gases. In 1784, guided by reports from other chemists who had collected the inflammable air given off when water reacts with iron, he conducted, with the aid of the young engineer J. B. Meusnier, a series of experiments that demonstrated water can be analyzed into oxygen and inflammable air.[68]

In the meantime the physicist J. A. C. Charles had, in a series of daring ascensions, demonstrated that inflammable air is an effective and manageable lifting agent for balloons. The construction of balloons thus created a practical demand for inflammable air. Lavoisier and Meusnier attempted to meet this demand by scaling up their apparatus for the analysis of water, and in 1785 they performed dramatic experiments on the production and collection of inflammable air.[69] The happy conjunction of truth and utility in eighteenth-century science had once again been demonstrated; the benefits to be realized by linking physics, chemistry, and aviation were made manifest.

7

The Campaign for French Chemistry

In 1787 Lavoisier marked the arrival of spring, as he had so often in preceding years, by delivering a public lecture at the Easter meeting of the Academy of Sciences. His text on this occasion was titled "The need to reform and improve chemical nomenclature." The subtext of this memoir is a masterfully constructed argument for the acceptance of his new theories. Ever the good general, Lavoisier assembled impressive forces before declaring his intentions. The memoir he read that April ignored the inadequacies of the phlogiston theory; so far as he was concerned, he had settled that issue two years earlier in his "Reflections on phlogiston." The memoir on chemical nomenclature marked the beginning of a new strategy that Lavoisier and his allies had recently formulated. They had decided to make reforming the language of chemistry the focal point of an ambitious effort to convince, and to a degree compel, all chemists to adopt Lavoisier's new theories and the methods on which they were based. The old chemistry was to be purged of its fundamental errors and the new theory of combustion was to be treated as fully verified. Lavoisier and his collaborators thus set out to alter radically the way chemists thought of and spoke about their science.[1]

Lavoisier began his memoir by acknowledging that he was reporting on work undertaken with Guyton de Morveau, C. L. Berthollet, and A. F. Fourcroy. They had worked closely together and had often been assisted by mathematicians in the Academy and by other chemists – he was doubtless referring to Laplace and Monge, among others. He then briefly summarized earlier revisions of the language of chemistry, commending in particular the work of Macquer and Baumé. This is interesting, for two years earlier, in his "Reflections on phlogiston", Lavoisier had directed withering criticism at the reformulations of the phlogiston theory proposed by these two academic colleagues. Macquer had died early in 1784, but his protégé Baumé remained a leading opponent of the new chemistry. The awkward truth was that as late as 1787 a majority of Lavoisier's fellow chemists in the Academy still refused to accept his new theories. As Lavoisier no doubt anticipated, Baumé was appointed to the four-man committee charged with determining whether the Academy should authorize publication of the new nomenclature. Another member of that committee, B. G. Sage, was also implacably opposed to the new chemistry; in fact, only one member of the committee, Cadet de Vaux, eventually joined Lavoisier's ranks.[2] Perhaps Lavoisier's gracious acknowledgement of Baumé's earlier contributions helped persuade the committee to recommend, after much huffing and puffing about the relative merits of the old and the new theories, that the Academy allow the new nomenclature to be placed before the public.[3]

Having placated Baumé with collegial courtesy, Lavoisier turned next to the early proposals of Guyton de Morveau, who had long been promoting the reform of chemical nomenclature.[4] In 1780 Guyton had translated a collection of essays by the prominent Swedish chemist, Torbern Bergman, and had been deeply impressed by Bergman's proposals for revising the language of chemistry.[5] At the same time he had agreed to write the articles on chemistry for an ambitious new project, the *Encyclopédie méthodique*. According to Lavoisier, accepting this responsibility made Guyton a spokesman for French chemistry, a duty he began to fulfill in 1782 when he published his initial proposals for reforming chemical nomenclature.

Destined to serve in some ways as the voice of French chemistry in a national publication, Guyton realized that in addition to creating a language, he had to ensure that it would be adopted, for the meaning of new terms can only be fixed by convention. He knew that before beginning the difficult assignment he had accepted, he needed to consult the chemists of France to find out what they thought his guiding principles should be, to show them the tables he planned to adopt for formulating a methodical nomenclature, and to ask them for at least their tacit consent. His essay was published . . . and he had the modesty to ask not for approbation, but for the critical comments of all those who cultivate chemistry.[6]

Lavoisier's description of Guyton's work was designed to promote a sense of national community among French chemists, as well as an awareness that the new nomenclature would only succeed if accepted by that community. On the first point he and Guyton were evidently of one mind. In the article on nomenclature Guyton wrote for the *Encyclopédie méthodique*, he explained with obvious pride why the reformers built their new language on Greek foundations: "In borrowing our terms from the language of a civilized people, whom the French have at all times most strongly resembled, and in ensuring that chemistry speaks the language of Aristotle and Plato, we provide the mind with a fertile source of suggestive associations."[7] Latin is the language of the Church; chemists look to the Greek philosophical heritage. This new language of science will omit all names derived from the myths preserved in barbarous tongues. It will provide instead a set of names accessible to all those whose education has prepared them to be citizens in the republic of learning. To understand the French terms formed from Greek roots, one needs "only the slightest and most introductory knowledge of the principal Greek words, a knowledge of the sort that every Frenchman who wishes to devote himself to a life of study and to the cultivation of the sciences acquires as part of his liberal education."[8]

Lavoisier, like Guyton, understood that language is a system of signs sanctioned by convention, and he too saw no reason to privilege natural or customary languages. The task of constructing new, rational languages was best left to experts. The new language of chemistry would, of course, incorporate the theories that science uses to explain natural phenomena; Lavoisier and his collaborators did not pretend that the new language they were constructing was theory neutral. The adequacy and appropriateness of the new language would be evaluated by the most eminent practitioners of the science. Only after the new language had been properly constructed and certified would the general run of chemists be invited to assist in its adoption by voicing their consent. This was the plan Lavoisier laid out in a series of carefully crafted paragraphs placed near the beginning of his memoir.

> Guyton correctly perceived that in a science that is in a state of flux, that is being rapidly improved, and in which new theories are being developed, it is extremely difficult to construct a language that fits different systems and that satisfies all views without adopting any one of them.
>
> To reassure himself in his research, Mr de Morveau consulted several chemists in the academy. This year he came to Paris for this purpose and offered to sacrifice his own ideas and his own research; he subordinated his desire for literary distinction to his love of science. In the meetings we held we all tried to act in the same spirit. We forgot what had been done, as well as what each of us had accomplished, so we could concentrate on what needed to be done. Only after reviewing all aspects of chemistry several times and after reflecting deeply on the metaphysics of language and on the relationship between ideas and words did we dare form a plan.[9]

This is extraordinary language for someone of Lavoisier's eminence to be using at the beginning of a formal public lecture in the Academy of Sciences. The religious allusions to self-sacrifice and willfully emptying the mind of both memory and self-interest are highly unusual, although their rhetorical purpose is clear. Guyton, the man with a mission to convert all French chemists to the new doctrines, had come to the high temple to be strengthened in his resolve and to be made wise. He was received with the respect due to a man of spirit who has had a vision, and the guidance he received was conveyed not as knowledge asserted to be true by a privileged group of academics, but as the fruit of deep reflection by a community of selfless priests.

What had in fact happened, as we shall see shortly, is that early in 1787 Lavoisier and his friends had convinced Guyton to abandon the phlogiston theory. Lavoisier's quasi-mystical account of Guyton's conversion was designed to persuade others to open their minds and accept the faith of the new brotherhood. By deploying the rhetoric of conversion in a public account of scientific progress, Lavoisier sought to invoke powerful modes of persuasion that were deeply embedded in Freemasonry, Christianity, and republicanism. If such cultural resources could be deployed in an academic lecture on the language of chemistry, we should not be surprised to find that comparable if more extreme appeals to moral fervor and national regeneration were invoked with Pentecostal force and suddenness at the very outset of the political revolution.

After laying out this ambitious strategy, Lavoisier assured the members of the Academy they would be shown the deference to which they were entitled. Two points were central here. First, although the new nomenclature was of pivotal importance to one of the sciences governed by the Academy, the proposal to recast the language of chemistry had not arisen within the Academy itself. It was therefore essential that the members of the Academy be given an opportunity to review the project and at the very least acknowledge its existence. Second, men of science and letters were responsible, as d'Alembert had written, to "fix the use of language" and "legislate for the rest of the nation in matters of philosophy and taste."[10] The Académie Française, the oldest and most exalted of the royal academies, had been invested with the authority to decide on questions concerning the French language. The members of the Academy of Sciences clearly had comparable authority over the subjects under their jurisdiction. Lavoisier left these points unstated, but they must have been in everyone's mind as he brought his introductory remarks to a close.

> We will limit ourselves for the present to describing our guiding assumptions and metaphysical principles. Once these principles have been stated, all that remains is to apply them, present the tables, and add summary explications. We will leave the tables on display in the Academy's meeting room as long as is thought necessary so that everyone can become thoroughly familiar with them and so that we can collect your reactions and refine our proposals through discussion.[11]

In the first part of his memoir Lavoisier related the new nomenclature to the concerns of the chemists in the Academy of Sciences; in the second part he turned to what he called the metaphysics of language. Lavoisier gratefully acknowledged that on this subject he was guided by the doctrines of the abbé de Condillac, whose well-known text on logic had been published in 1780.[12] Condillac's philosophy of language, while striking, was not considered especially daring in Paris in the

1780s and Lavoisier and his colleagues were taking no risks in stressing its applicability to science. Language, Lavoisier said, is an analytic method; indeed algebra itself is a language. Language is an instrument used by the mind and, like all instruments, it ought to be as well constructed as possible. Language is composed of signs that represent ideas that arise in the mind when it encounters the facts of nature. Children learn language naturally because immediate experiences of pleasure and pain tell them when they have and have not formed proper ideas about nature. Because this natural system of verification is absent in science, many investigators cling to unwarranted prejudices. Scientists must therefore constantly check their reasoning against experiments and observations. The experimental method is thus the key to the construction of a rational language, for a scientific language should embody the logic of science. The language of science should also faithfully reflect reality.

When two years later Lavoisier again summarized Condillac's philosophy of language in the preface to the textbook he wrote for those who were just beginning the study of chemistry, he again declared that Condillac's account of language justified his dramatic recasting of chemistry's theories and nomenclature. Those already predisposed toward Condillac's philosophy may have been persuaded by this argument for the new language of chemistry. Historically, however, it is quite clear that before 1783, that is to say when Lavoisier was constructing his new theories, the way he engaged conceptual and theoretical problems was not fundamentally shaped by a commitment to a particular philosophy of language. Lavoisier adhered to the method of experimental physics during the years of theory construction because he had been persuaded to do so by Lacaille and Nollet, not because Condillac's theory of knowledge encourages reliance on experiment. His new theory of combustion and the new nomenclature were rooted in the chemistry he learned while studying with La Planche and Rouelle; Condillac's account of the relations between facts, ideas, and words merely provided Lavoisier and his collaborators with a convenient way to package and present their conclusions. Lavoisier campaigned vigorously for a radical transformation of the language of chemistry because doing so gave him an opportunity to impose his new theories first on those who practiced chemistry in France and then, he hoped, on chemists everywhere. Condillac provided a convenient horse that Lavoisier rode hard; the origins and value of the goods he carried in no way depended on the strength of the mount.

After briefly reviewing the older language of chemistry, which he predictably characterized as full of confusion and error, Lavoisier summarized the principles that he and his collaborators employed when creating their new system of names. He left the details on this matter to the longer memoir that Guyton read at a closed meeting of the Academy two weeks later. Lavoisier restricted himself to providing just enough information to indicate how tightly the new nomenclature was linked to his theories. He emphasized that he and his colleagues were proposing "a method for naming rather than a list of names."[13] This privileging of the systematic spirit over the spirit of system eliminated any need to begin with an ontological theory of elements. In the new nomenclature simple substances, which functioned as elements, were defined as substances that had not yet been analyzed. When naming such substances one should focus on their most general and distinctive properties; when naming compounds one should adhere to the Linnaean binomial principle of providing two terms, the first generic, the second specific.

Lavoisier then illustrated how these principles were to be applied.

> The acids, for instance, are composed of two substances that we consider simple, one of which conveys acidity and is found in all acids. This is the substance from which the generic name is taken. The other substance is specific to each acid and is different in each, indeed it gives rise to the differences among the acids. This is the substance from which the specific name is taken.[14]

He then applied these rules to the naming of metallic calxes and other combustible substances. In each case he assumed, as he had when constructing his theories, that there is one generic substance that causes acidity and combines with bodies when they are calcined or burned. In this way the pattern of explanation that was fundamental to his research program was made central to the new nomenclature. Chemical phenomena were to be explained by the enduring chemical properties of certain key constituents. Once this fact is recognized, it is no longer surprising that Buffon and others, who pursued different chemical research programs, rejected the new nomenclature. Indeed, since the new nomenclature was by design and use a key weapon in the campaign against both Newtonian chemistry and the phlogiston theory, one can only wonder why its introduction was not more stoutly resisted than it was.[15]

The ties that bound Lavoisier, Guyton, Fourcroy, and Berthollet together in the campaign for the new nomenclature were not as simple as one might expect. They were united in their rejection of the phlogiston theory, yet they were not of one mind on many other aspects of the new chemistry. They were joined in a diplomatic alliance rather than being believers who had all converted to a common faith. One did not have to accept all of Lavoisier's theoretical claims to join his campaign, and in fact few of his allies did so.[16]

Lavoisier's theories evolved over the course of many years, and while he succeeded in generalizing them into an integrated system that accounted for nearly all the phenomena previously explained by the phlogiston theory, his contemporaries were free to pick and choose which parts of this system they believed to be true. Lavoisier had no doubts about the truth of his theories, yet he showed unfailing patience when responding to objections raised by competent chemists who remained unpersuaded. He was an effective leader precisely because he always marshalled just the evidence, arguments, and allies needed for the particular contest at hand. He understood that true scientists must be persuaded rather than compelled and he showed consummate skill in making use of all the considerable means of persuasion available to him.

Claude Louis Berthollet was the earliest in this group to join Lavoisier in abandoning phlogiston, but he never did accept his theory of acidity.[17] Like both the phlogistonists and the antiphlogistonists, Berthollet believed that the properties of compounds are caused by the enduring chemical properties of general principles. In the mid-1770s he was investigating the nature of air and hoped to find a universal acid, yet he rejected Lavoisier's interpretations of calcination and combustion while remaining firmly in the phlogistonist camp. In the latter part of the decade he was increasingly impressed by Lavoisier's experiments, but when elected to the chemistry section of the Academy in 1780, he still considered sulfur a compound and vitriolic acid simple.

As the phlogistonists revised their doctrines in response to Lavoisier's relentless attack, Berthollet sided with Macquer, Kirwan, and others in arguing that inflammable air is pure phlogiston. But when Lavoisier demonstrated that oxygen and inflammable air combine to form water, Berthollet concluded there was no further reason to believe that phlogiston exists. Lavoisier had at last won over someone whose opinion carried great weight in the community of French chemists. Berthollet's acceptance of Lavoisier's basic orientation proved to be especially valuable in the long run, for in the early years of the nineteenth century he and Laplace kept alive the tradition of experimental research that Lavoisier had established and directed until shortly before his death.[18]

Fourcroy slipped into alliance with Lavoisier so gradually that we cannot say just when they found themselves on the same side of the theoretical fence.[19] Twelve years younger than Lavoisier and a popular lecturer and author of textbooks, Fourcroy was particularly well-equipped to help the antiphlogistonists carry their case to the larger public. When Macquer died in February 1784, he left vacant two important posts, the chemistry professorship at the King's Garden and the directorship of the Gobelin dye-works. Fourcroy and Berthollet were the two leading candidates for the professorship and each naturally had a patron. The Duke of Orléans supported Berthollet, who was already a member of the Academy, but Buffon, the director of the King's Garden, favored the younger and more experienced lecturer. Buffon prevailed and Berthollet was appointed to the Gobelin position, and in the end both appointments proved to be successful. It must have given Lavoisier considerable pleasure when a few months later Buffon's protégé and Macquer's successor jettisoned phlogiston and joined those campaigning for the new nomenclature.

Sometime between the latter part of 1784 and the middle of 1786 Fourcroy abandoned Macquer's revised version of the phlogiston theory and began teaching Lavoisier's theories in his immensely popular public lecture course at the King's Garden. His acceptance of Lavoisier's position on phlogiston may have been related to his candidacy for election to the Academy. In 1785, while serving as director of the Academy, Lavoisier succeeded in reorganizing the sections along lines he had proposed over twenty years earlier. Two new classes were created, one in natural history and mineralogy and one in general physics, and the total number of members was increased slightly. The reassignments that followed created a vacancy in the chemistry section, to which Fourcroy was promptly elected.[20]

It is hard to imagine that Fourcroy had not already indicated where he stood on theoretical issues before being elected, for at that point Berthollet was Lavoisier's only ally in the chemistry section. In any case it was not long before Fourcroy publicly acknowledged his shift in the introduction to the second edition of his widely read textbook *Elements of Natural History and Chemistry* (*Élémens d'histoire naturelle et de chimie*). The third edition, expanded to five volumes, was issued late in 1788 and made full use of the new nomenclature. Lavoisier could not have asked for a more effective disciple and Fourcroy deserves much of the credit for the rapid dissemination of the new chemistry.[21] It was probably the success of Fourcroy's texts that persuaded Lavoisier to abandon his own projected general textbook and led him to focus instead on preparing a shorter thematic introduction to the new chemistry.

Fourcroy, like Berthollet, also played a major role in preserving and shaping the historical memory of Lavoisier's scientific methods after the Terror had run its course. In 1790, when Guyton was obliged to give up his duties as author of the chemistry volumes for the *Encyclopédie méthodique*, Fourcroy agreed to take on this task. In volume III, published in 1796 or 1797,[22] he provided a memorable description of the scientific community that gathered in Lavoisier's apartment and laboratory at the Arsenal.

> Twice a week at his home he held gatherings to which were invited those men most distinguished in geometry, physics, and chemistry; instructive conversations, exchanges resembling those which had preceded the establishment of the academies, there became the center of all enlightenment. There were discussed the opinions of all the most enlightened men of Europe; there were read the most striking and the newest passages from works published by our neighbors; there were theories compared with experiment . . . I shall never in my life forget the privileged hours which I spent in those erudite exchanges where it was so pleasant for me

to be admitted ... Among the great advantages of those meetings, that which struck me the most and whose invaluable influence soon made itself felt in the heart of the Academy of Sciences, and subsequently in all the works of physics and chemistry published for twenty years now in France, is the harmony which was established between the manner of reasoning of the geometers and that of the physicists. Precision, severity of the language, of the expressions, and of the philosophical method of the former, passed gradually into the minds of the latter.[23]

Guyton de Morveau was the last of Lavoisier's three key allies to join the campaign.[24] In 1786, after completing the initial part of his first volume for the *Encyclopédie méthodique*, Guyton began working on an article on "Air." Already deeply concerned about nomenclature, he realized that the terms he used to describe the airs would depend on the theory of air he adopted. This, of course, was a highly contested area of science and so he went to Paris for guidance. He arrived in February 1787 and in the end stayed for seven months.

It evidently did not take Lavoisier and his friends long to bring Guyton around to their way of thinking, for in March Lavoisier wrote to Meusnier that "Mr de Morveau is now in Paris and we are taking advantage of this opportunity to work with him on a chemical nomenclature. This is arguably what is now most needed for the advancement of the sciences."[25] Lavoisier, delighted to have won over the most prominent protégé of both Buffon and Macquer, was now prepared to argue more openly for the new chemistry. While composing his introductory essay for the Easter meeting he urged Guyton to complete his memoir on the principles to be followed in naming chemical substances and enlisted Fourcroy to prepare a large table that would set the old and the new names side-by-side in six columns. When all this had been done, Lavoisier was ready to begin a new phase in his long campaign for hegemony over French chemistry.

An interval of four years separated Lavoisier's and Laplace's "Memoir on heat" from Lavoisier's Easter memoir on the need to reform the language of chemistry. Midway between these two presentations Lavoisier read what was to become one of the most famous texts in the history of chemistry, his "Reflections on phlogiston." This rhetorical tour de force has traditionally been considered his final word on the phlogiston theory, and that is certainly what he intended it to be. Historically, however, this memoir, despite its ferocious brilliance, did not succeed in routing the old chemistry and installing the new.[26] Lavoisier may have found this absence of impact puzzling, but within a year he evidently concluded that a different strategy was called for. It was at that point that he began campaigning for a new language for chemistry.

The full title of the memoir Lavoisier read on 28 June and 13 July 1785 is "Reflections on phlogiston, a sequel to the theory of combustion and calcination published in 1777."[27] As noted in the previous chapter, Lavoisier composed his 1777 memoir on combustion before he and Laplace had systematically examined the flow of heat in changes of state and chemical reactions. In the 1777 paper he acknowledged that his explanation of combustion was a theory formed by generalizing from four experimentally determined laws. He argued in the earlier paper that the supposition that phlogiston plays a role was no longer necessary, but he also restricted himself to claiming that his new theory should be accepted because it was better than the phlogiston theory, not because it was demonstrably true. In 1785 he went much further, boldly insisting that his fully elaborated theory of combustion is undeniably true and that all versions of the phlogiston theory are demonstrably false. The polite discourse of inductive generalization yielded to the deductive reasoning of the authoritative theorist. The difference in tone is striking.

The "Reflections on phlogiston," when seen in the context of Lavoisier's abiding commitment to the experimental method, appears anomalous. Composed at the end of a period of intense theory construction, it represents Lavoisier's thinking at a moment when he was evidently so impressed by the power and range of his novel insights that he temporarily threw aside the methodological caution he had exhibited throughout his career and wildly overreached himself. Whereas the earlier memoir on combustion sought to persuade, the "Reflections on phlogiston" seek to command. Lavoisier offers an alternative to phlogiston that is hypothetico-deductive in structure. His strategy was to force the reader to choose between two alternative theories he offered as the only possible choices. Both papers were read to closed meetings of the Academy; the latter was distinctive in that it attempted to score a knockout, a *coup de science*.

For a brief interlude Lavoisier, the Academy's foremost chemist and its director in 1785, embraced the spirit of system. His colleagues naturally resisted his frontal attack, for while they acknowledged the force of his indictment of phlogiston, they could not and would not be compelled to accept the theories he offered to replace it. Perhaps Lavoisier, when he realized that not even his closest allies were prepared to accept all his theoretical claims, recognized the rashness of his grab for power and the legitimacy of their skepticism. After recovering his poise as an experimentalist and positivist and forsaking the siren song of deductive systems, he changed direction again and set out, with the help of his collaborators, to win the victory he sought by capturing control over the discourse of chemistry.

The "Reflections on phlogiston," like Lavoisier's other memoirs, begin with carefully phrased methodological observations that set the tone for what follows. After listing several topics he had addressed in earlier memoirs, Lavoisier claimed they could all be explained by invoking the oxygen principle and the substantial fire with which it combines. "Once this principle is admitted," he asserted, "the main difficulties of chemistry appear to evaporate, and all phenomena are explained with astonishing simplicity."[28] This promise of completeness and simplicity was then turned against the phlogiston theory.

> If all of chemistry can be explained in a satisfactory manner without the help of phlogiston, that is enough to render it infinitely likely that the principle does not exist, that it is a hypothetical substance, a gratuitous supposition. It is, after all, a principle of logic not to multiply entities unnecessarily. Perhaps I should provide falsifying arguments and be satisfied with having proved that one can account for these phenomena better without phlogiston than with it. But the time has come for me to examine precisely and formally a view that I consider a profound error in chemistry, one that has introduced a type of reasoning that has, I believe, seriously retarded its progress.[29]

Then, like others who have sought to proceed by manipulating ideas rather than by referring to or expanding the stock of empirical knowledge, Lavoisier asked his colleagues to empty their minds.

> At the beginning of this memoir I ask my readers to discard, in so far as is possible, all prejudices, to see the facts presented to them for what they are, to banish all suppositions based on reason, to transport themselves

back to the pre-Stahlian era, and to forget for a moment, if it is possible, that his theory ever existed.[30]

Lavoisier's immediate purpose was to rescue through reinterpretation, rather than to discredit, G. F. Stahl, the original author of the phlogiston theory. He was, Lavoisier insisted, "one of the patriarchs of chemistry who revolutionized the science to a certain degree." Stahl, according to Lavoisier, could only understand combustion in terms of "what struck his senses," that is to say the release of light and heat. These perceptions led him to suppose that an inflammable principle is the active agent in combustion. Although his disciples then speculated about the nature of this substance, Stahl himself should be remembered for having made two key observations, observations Lavoisier called "eternal truths." The first is that the calcination of metals is a form of combustion, the second and more important is that the property of inflammability can be transferred from one substance to another. Lavoisier provided an example of this second observation by pointing out that when the combustible substance charcoal is mixed with the non-combustible substance vitriolic acid, the sulfur in the acid recovers its combustibility.

Stahl, according to the reconstruction of history that Lavoisier wished to inscribe on the blank slate of his readers' minds, was like the innocent child of Condillac's theory of learning in that he could only note and reason about information presented directly to his senses. But his successors in the post-Edenic world of science had been taught how to perform experiments and why they must do so, and their failure to remedy the flaws of Stahl's theory could not be so easily excused. For the fact is, as had been known since the seventeenth century, substances gain weight when they burn. Thus if phlogiston is lost during combustion, it must have a negative weight. It was this feature of the theories of Stahl's successors, and not Stahl's original supposition that there is an inflammable principle involved in combustion, that Lavoisier subjected to relentless criticism in the "Reflections."

We need not linger over the fun Lavoisier had dismembering the theories of his contemporaries. The writing is brilliant, the contradictions exposed are numerous, the tone mocking but not overtly insulting. Priestley, Macquer, and Baumé are shown to have held views that cannot be made compatible with one another or with well-known experimental facts. Lavoisier indirectly recalled Newton's crucial experiment on the composition of white light when he reminded the phlogistonists, who used their theory to explain the origins of colors in bodies, that whiteness, rather than being an absence of color, is a mixture of all the colors.[31] When the fun ended halfway through the memoir, Lavoisier summarized his case against phlogiston in two ringing paragraphs that chemists still cite when recalling the fall of the old regime.

> These reflections confirm what I have stated, wish to prove, and will repeat again: that chemists have made phlogiston a vague principle that is not rigorously defined and hence can be adapted to whatever explanations one wishes. Sometimes the principle has weight and sometimes it does not, sometimes it is free fire and sometimes fire combined with an earthy element, sometimes it passes through the walls of containers and sometimes it cannot penetrate them. It explains simultaneously causticity and non-causticity, transparency and opacity, colors and the absence of color. It is a veritable Proteus which changes form instantaneously.
>
> It is time to lead chemistry to a more rigorous way of reasoning, to strip the facts with which this science is enriched every day of whatever speculation and prejudice have added to them, to distinguish facts and observations from systems and hypotheses, and finally to take note of the current state of chemical understanding so that those who succeed us can move forward with confidence.[32]

In the latter half of the memoir Lavoisier put forward an elaborate theory of heat. Surprisingly, he did not begin with quantitative experiments, as he and Laplace had in their "Memoir on heat," but rather with highly hypothetical transduction. When bodies are heated they expand; when they cool they contract.

> From this we see that the molecules of bodies do not touch, that there is a space between them that heat increases and cold diminishes. One can hardly think about these phenomena without admitting the existence of a special fluid which when accumulated causes heat and when rarefied causes cold. It is no doubt this fluid which gets between the particles of bodies, separates them, and occupies the spaces between them. Like a great many physicists I call this fluid, whatever it is, *the igneous fluid, the matter of heat and fire*.[33]

This is hypothetical reasoning worthy of a Descartes. Lavoisier pressed on undaunted, however, and proceeded to explain changes of state by supposing that there are three forces involved, the force of attraction, the expansive force of fire, and the weight of the atmosphere. It is the atmosphere that gives rise to the liquid state, for if there were no atmospheric pressure solids would become gases as soon as the elastic force of the heat they contain overcame the attractive force holding their molecules together.

Having stated his assumptions, Lavoisier proceeded to a thought experiment as abstracted from reality as any of Galileo's famous thought experiments in kinematics. His purpose was to build up an idealized picture of the interaction of heat and molecular matter. His guiding principle was taken from Adair Crawford's book on heat, which he cited; Crawford in turn had taken his central concept from the Scottish chemist, William Irvine.[34]

Irvine's theory, about which Joseph Black had grave doubts, attempted to link latent heat, the heat absorbed or given off during changes of state, to specific heat, the heat involved in temperature changes in bodies in a given state. According to Irvine and Crawford, the greater the latent heat needed to separate the molecules so as to cause a change of state, the greater the specific heat of the substance. As Lavoisier put it, in a most unconvincing way, "It is clear a priori and independent of all hypotheses that the greater the distance between the molecules of bodies, the greater their capacity to receive the matter of heat, and consequently their specific heat will be greater."[35] It is not surprising that Lavoisier found Crawford's theory attractive; what is surprising is his willingness to present it in a hypothetical form that depends upon a series of suppositions about the microscopic nature of matter and heat. The 1783 "Memoir on heat" clearly indicates Lavoisier had been thinking about heat in these terms for several years. Normally, however, he was careful not to claim in public that the models he used to organize his thinking were truths about nature. Priestley must have chuckled when reading this philosophical epistle from the Parisians.

Lavoisier plunged on, so ravished by his theory of heat that he completely ignored his theory of acidity, the fulcrum on which, according to his earlier account, his theory of combustion turned. Perhaps

Berthollet's and Fourcroy's reservations about the universality of his theory of acidity had given him pause. What we see in the "Reflections" is the confident theorist in full cry. The questions being addressed were both chemical and physical, and Lavoisier once again invoked the analogy of salt crystallization,[36] but the alternative he offered to the discredited phlogiston theory was more physical than chemical. How the chemists must have winced when he said that "in so far as the matter of heat is concerned, nearly all natural bodies are like sponges in water."[37] If this was what came of working with Laplace, Monge, and Vandermonde, all of whom Lavoisier cited, it is hardly surprising that Lavoisier's chemical colleagues responded harshly to the case he made against the phlogiston theory.

The reaction was immediate. The Dutch chemist Martinus van Marum was visiting Paris and attended the meeting at which Lavoisier finished reading his "Reflections." He was astonished by the "violent objections . . . made against this [paper], as a result of which the reading was continually interrupted. This, together with the simultaneous efforts of the reader and of his opponents to be heard, led to my understanding very little."[38] Lavoisier may not have been surprised by this response, for he certainly was not innocent in these matters. His closing paragraph, which is often quoted as an abstract observation about scientific change, threw a final dash of salt on the chemists' wounds. One can only marvel at the brazenness with which Lavoisier taunted his fellow members of the chemical section.

> I do not imagine that my ideas will be adopted all at once. The human mind becomes accustomed to particular ways of seeing things and those who have perceived nature from a certain point of view during part of their careers will find it difficult to adjust to new ideas. Time alone will confirm or destroy the opinions I have presented. In the meantime, I note with great satisfaction that young men who are beginning to study science without any prejudices, and mathematicians and physicists who are open-minded about the truths of chemistry, no longer believe in phlogiston in the sense in which Stahl presented it. They consider this doctrine a scaffold that is more embarrassing than helpful for the further construction of chemical science.[39]

What Lavoisier had forgotten in 1785, or had willfully suppressed, was that while arguments that invoke crucial experiments may appear to force one to choose between two options, other responses are in fact possible. One may agree that one of the proposed theories has been falsified and the alternative theory has been confirmed, but one can also conclude that neither of the available theories is well grounded and that the question remains open. The "Reflections on phlogiston" was essentially Lavoisier's last word in the phlogiston debate; after that it was enough to reiterate the list of contradictions that plagued the theory. But conjecture and refutation were not tightly linked in eighteenth-century chemistry and some other way had to be found to move chemists from a state of philosophic doubt about the problems of combustion, calcination, and acidity to acceptance of his new theories. All this became clear between 1785 and 1787. It was during this period that Lavoisier backed away from his colleagues in physics and began working more closely with his new allies in chemistry. With their help he devised a broad-based strategy designed to win chemists over to his new theories. The new nomenclature was their primary *machine de guerre*, but they also employed a variety of well coordinated secondary weapons as well.

The campaign for the new chemistry in France, and in other nations as well, was prolonged and intensely contested.[40] The most determined and effective of Lavoisier's French opponents was Jean-Claude de Lamétherie.[41] In 1785 Lamétherie became editor of Rozier's *Journal de Physique*, the foremost Parisian outlet for rapid publication of new scientific information. In his annual surveys of the sciences and in notes added to the articles selected for publication, Lamétherie found ample opportunity to comment on the Lavoisians' campaign. At first his criticism was temperate, but as he became more isolated in Paris and as the circle of foreign correspondents opposed to the new chemistry grew, he became increasingly contentious and outspoken. Like several of his French allies, Lamétherie was driven to distraction by the condescension of the academicians, who never deigned to admit him to their ranks.

Lamétherie's control of the *Journal de Physique* created a real problem for the antiphlogistonists and they soon realized they would have to find another outlet for their work. As early as 1787 they considered publishing a French version of Crell's *Chemische Annalen*; two years later they formally presented the first volume of their new journal, the *Annales de Chimie*, to the Academy. The editorial board of the new journal included Lavoisier, Guyton, Berthollet, Fourcroy, and Monge, as well as three additional recruits to the cause, the Baron de Dietrich, J. H. Hassenfratz, and P. A. Adet. This new chemical journal, founded just as the political revolution was beginning, proved invaluable in the campaign against Lamétherie and the coalition he headed.[42]

The *Method of Chemical Nomenclature* (*Méthode de nomenclature chimique*), the book that introduced the new language of chemistry to a wider public, was published in the summer of 1787. It contains Lavoisier's introductory memoir, Guyton's account of the principles to be followed when forming new names, Fourcroy's table of the old and new names for simple substances and selected compounds, and two long appendices, one a listing of synonyms and the other a dictionary.[43] At nearly the same time Richard Kirwan's *Essay on Phlogiston* appeared in England. Kirwan, like Macquer, argued that inflammable air is phlogiston. His calmly reasoned defense of his beliefs gave the antiphlogistonists an opportunity to address British arguments in favor of phlogiston without having to tangle with Priestley. Madame Lavoisier undertook the task of translating Kirwan's text into French and all the leading allies, that is to say Lavoisier, Guyton, Berthollet, Fourcroy, Laplace, and Monge, composed lengthy notes answering Kirwan's arguments. The annotated French translation was published in 1788; a year later a second English edition of Kirwan's book, along with the French notes and the author's replies, was published in London.[44] This informed and courteous exchange of views eventually led to a rational close. In 1791 Kirwan abandoned phlogiston, but he still could not bring himself to accept fully the new nomenclature championed by his French contemporaries.

Less visible but of great importance were two other aspects of the campaign for the new chemistry.[45] One of these was a concentrated effort to persuade visiting scientists on an individual basis to accept the new theories. All the members of the inner guard participated in these encounters and it cannot have been easy to resist the combined attentions of so many distinguished scientists when visiting the cultural capital of Europe. Lavoisier's authority was, of course, pre-eminent and he deployed it with tact and considerable effect. The Dutch chemist van

Marum was given the full treatment during his visit to Paris in 1785; by 1787 he had brought out a work in Dutch supporting Lavoisier's views. When the young Scot, James Hall, a well-informed advocate of Joseph Black's chemistry, paid a lengthy visit to Paris in 1786 while on a grand tour of the Continent, he was invited to dine with Lavoisier on Mondays. Lavoisier, whom Hall reported "has one of the clearest heads I have ever met with," flattered Hall and his mentor by acknowledging in carefully chosen words that "the whole [combustion theory] is founded on the theory of latent heat" and that "latent Heat and fix[ed] air [carbon dioxide] are two of its foundation stones."[46] Disarmed and persuaded, Hall accepted Lavoisier's conclusions; after returning to Scotland he became a key advocate for the new chemistry among the scientists of Edinburgh.

The other, less visible tactic employed by the antiphlogistonists was to write letters refuting the arguments of those who attacked the new chemistry and reinforcing the faith among those who had accepted it. Although scientific publication was rapid and voluminous in the eighteenth century, correspondence networks among scientists continued to flourish, as they had in the seventeenth and as they still do late in the twentieth century. One can say things in a letter that one would not wish to see in print, and many did. The quantity, rapidity, and importance of such correspondence in the eighteenth century, like the frequency and importance of personal visits among scientists, comes as a surprise to those who know only the published record of science. Lavoisier, who could afford to devote only a small part of his time to science in the late 1780s, did not play a significant role in these correspondence networks, but his allies in Paris and elsewhere became increasingly active in his behalf as the campaign heated up. And as was the case with personal persuasion, the aura of authority that surrounded those who occupied the high ground of the Academy of Sciences lent great weight to the reports that radiated out via the post to interested chemists in Europe and America.

The capstone of the campaign to achieve hegemony over French chemistry was Lavoisier's classic *Elementary Treatise on Chemistry*.[47] Unlike many other aspects of the campaign, the composition of this text was not a collaborative effort, and its publication early in 1789 marked both the end of the closely coordinated allied campaign and Lavoisier's reassertion of his proprietary claims as the author of the new chemistry. The *Treatise* brought the chemical revolution to a triumphal conclusion. Thereafter, when Lavoisier had time to pursue chemistry, he turned his attention to extending his research program in physiological and organic chemistry.

Lavoisier began the preface to the *Treatise* by recalling his earlier memoir on nomenclature.

> When I began the following Work, my only object was to extend and explain more fully the Memoir which I read at the public meeting of the Academy of Sciences in the month of April 1787, on the necessity of reforming and completing the Nomenclature of Chemistry.[48]

He devoted the remainder of the preface to a re-exposition of Condillac's philosophy of language and its applicability to chemistry. His argument, which essentially repeats what he had said earlier, is a classic in the pedagogical literature of the Enlightenment. Those who had the good fortune to begin their study of chemistry with Lavoisier's *Treatise* were spared the confusion of names and suppositions that had so befuddled his own first encounter with the subject. Lavoisier purged chemistry of a pervasive but erroneous set of ideas and replaced them with a comprehensive new theory; he transformed chemistry into a true science by insisting it abide by the method of experimental physics and by providing it with a rational language. His goal in the *Treatise* was to make smooth the way for beginners who wished to undertake the study of the science to which he had devoted his career.

The fluency and self-confidence of the *Treatise* are commanding. As is generally true of Lavoisier's public presentations, however, they are features of the rhetorical mask he donned after resolving problems that at first seemed intractable. Lavoisier's achievements did not come easily, but, unlike Jean Jacques Rousseau, he saw no reason to burden his audience with accounts of the dark nights in which his soul was sorely tried by doubt, confusion, and frustration. His image of himself and the image he projected was that of a rational, cultivated, and polite public figure. It is an image that set the tone for almost all his scientific presentations.

Lavoisier had in fact spent years worrying over how to organize an introductory chemistry text.[49] Late in 1777, after reading his "Memoir on the general nature of combustion," he began sketching out a plan for a second volume of *Physical and Chemical Essays*, but nothing came of it. At about the same time his colleague J. B. Bucquet was tutoring Madame Lavoisier in chemistry and Lavoisier was apparently prompted by this effort to take up again the problem of how best to organize a text for students beginning to study the subject. The two men worked together on the project, but their collaboration ended suddenly when Bucquet died at the beginning of 1780. During the next two years Lavoisier sketched out two versions of their proposed text and drafted a "preliminary discourse on the application of logic to the physical sciences and to chemistry in particular." This prefatory essay includes many passages from Condillac's text on logic, which had just been published. But this project was left incomplete and Lavoisier filed it away when he turned his attention to experiments on heat.

Lavoisier returned to these earlier drafts in 1787 and 1788, and essentially all his later statements on the application of Condillac's theory of language to chemistry can be found in these earlier manuscripts. By 1787 all the elements of his new chemical theory were in place and the science had a new language, but it was still far from clear how an introductory text for the new chemistry should be organized. Comprehensive chemistry texts based on lecture courses offered one model. Although Rouelle's lectures were never published, full manuscript versions were widely available and provided an admirable example of this type of text. Such texts gave greater attention to discrete descriptions of chemical substances and processes than to theoretical integration, however, and we know that Lavoisier had been put off by this seriatim approach when he first began studying chemistry. Since he never delivered comprehensive lectures on chemistry, Lavoisier was probably happy to leave this demanding job, and the associated task of introducing theoretical rigor into texts derived from such presentations, to his colleagues. Fourcroy, who enjoyed great success as a lecturer and author of texts in the years following Bucquet's death, appeared to have this facet of the campaign well in hand by the time Lavoisier again took up the problem of how best to introduce students to the new chemistry.

The dictionary format was another model available to Lavoisier; Macquer had used it with great success. This way of organizing knowledge was especially popular in the middle of the century, when those opposing hierarchical systems of natural philosophy were attacking the notion that all knowledge can be arranged in rigorously deductive systems. But while breaking subjects down into discrete dictionary entries made possible a high degree of clarity and accessibility, it did nothing to remedy the problem of intellectual integrity. Lavoisier was happy to associate himself with encyclopedias and dictionaries when they eased the transition to the new nomenclature, but he believed that the explanations offered by scientists should be organized around real causes, not conventional terms. Fundamentally he was committed to theory rather than taxonomy.

In 1788 Lavoisier decided to emphasize clarity and method rather than completeness when drafting his *Elementary Treatise*. His goal, he wrote, was "to keep everything out of view which might draw aside the attention of the student," and hence he eschewed all history and all citation of other investigators. He organized his text into three sections. The first, based entirely on his own work, begins with the theory of caloric. Lavoisier decided to call the substance that causes heat caloric after rejecting the more logical term thermogen, i.e. the cause of heat.[50] He explained how caloric causes changes of state and combines with substances to form aeriform fluids (gases). The discussion then moves to the composition of the atmosphere. After describing oxygen gas, he presented the various ways it can be decomposed: by combining with such substances as sulfur and phosphorus to form acids, by combining with metals to form calxes, and by combining with hydrogen to form water. The remainder of the section is given over to discussions of particular examples of combustion, acidification, analysis, fermentation, acid-alkali neutralization, and the properties of bases and metals.

The second section examines the combination of acids and bases and the formation of neutral salts. Lavoisier forthrightly admitted that "this part contains nothing which I can call my own";[51] he included it primarily because he considered the experimental study of salts, to which Rouelle had made important contributions, the only aspect of earlier chemistry that was well-grounded. At the beginning of the section he provided a table of simple substances. A substance is to be considered simple when it cannot be analyzed further by chemical means: "The principle object of chemical experiments is to decompose natural bodies, so as separately to examine the different substances which enter into their composition."[52] As new instruments of analysis become available, the list of simple substances will change in ways that cannot be anticipated. By 1788 Lavoisier had evidently abandoned the hope, which informed his "Reflections on phlogiston," that chemistry could be transformed into a deductive science.

> As chemistry advances towards perfection, by dividing and subdividing, it is impossible to say where it is to end; and these things we at present suppose simple may soon be found quite otherwise. All we dare venture to affirm of any substance is that it must be considered as simple in the present state of our knowledge, and so far as chemical analysis has hitherto been able to show.[53]

The list of simple substances given in Lavoisier's *Treatise* is divided into four parts.[54] The first consists of "simple substances belonging to all the kingdoms of nature, which may be considered as the elements of bodies" and includes light, caloric, oxygen, azote (nitrogen), and hydrogen. These are the widely distributed chemical constituents of nature that at the end of the chemical revolution did the theoretical work that air, water, and fire had performed before. The second part of the list contains six acidifiable non-metallic substances, the first three being sulfur, phosphorus, and charcoal, the second three being substances about which Lavoisier said, "we only know that these radicals are susceptible of oxygenation, and of forming the muriatic, fluoric, and boracic acids."[55] The third part of the list contains seventeen "oxydable and acidifiable simple metallic bodies" ranging from antimony to zinc. The fourth part contains five "salifiable simple earthy substances," the first two being lime and magnesia. At an earlier date the substances contained in the latter three lists were all called earths.

The bulk of the second section of the *Treatise* consists of detailed discussions of the properties of the simple substances and their compounds. The third section is an extended "description of the instruments and operations of chemistry." Here and in the attached plates Lavoisier proudly displayed the vast array of expensive instruments he employed in his research. This was not simply an exercise in self-aggrandizement, however, for his purposes were didactic and methodological. "The method of performing experiments," he declared, "and

180

particularly those of modern chemistry, is not so generally known as it ought to be."[56] This section, like the first, was firmly based on his own investigations.[57] Lavoisier did not claim that he had helped formulate the methodological principles of experimental physics, but he did believe he had played a significant role in introducing them into chemistry. In a discussion of fermentation he concisely summarized the methodological postulates that were, in his opinion, fundamental to chemistry.

> Nothing is created either by human action or in natural operations. It is a fundamental truth that in all operations there is the same quantity of matter before and afterwards and that the quality and quantity of the material principles are the same; there are only alterations and modifications. The entire art of making chemical experiments is founded on this principle. One must assume in every case that there is a true equality or equivalence between the material principles of the bodies examined and those obtained by analysis.[58]

This statement mandated strict adherence to the methods of experimental physics; it also obliged one to refer to the qualitative properties of material principles when explaining the chemical properties of compounds. Both these methodological commitments were presuppositions Lavoisier brought to the study of specific chemical problems. They were in this sense "metachemical," and while not all chemists accepted them or abided by them, Lavoisier, from the beginning of his career to its end, never considered them problematic.

Publication of the *Treatise* did not lay to rest Lavoisier's urge to write a more comprehensive text for students of chemistry. In 1792, after spending several months revising his many memoirs and writing historical introductions for a projected multi-volume edition of his works, he turned once again to the preparation of what he called both "a course of experimental chemistry organized according to the natural order of ideas" and a text in "experimental philosophy."[59] The outline he drew up contains a short introduction on the relations between physics and chemistry and a long preliminary discourse on how chemistry should be taught and what subjects students should study before taking up chemistry.[60]

The plan of the proposed text was conceived in very broad terms. It begins with a discussion of the general properties of bodies and addresses general questions about the nature of matter and the laws of physics. An extended discussion of the physics of light follows and only then does Lavoisier begin to address the subjects examined in the *Treatise*. Once these topics had been covered, he proposed to add sections on chemical technologies, such as mining, glass-making, and saltpeter production, and on geology and mineralogy. There are indications he thought of this as a second edition of his *Treatise*, and had he lived to complete it, our conception of that text today would be very different from the one we now hold.

The campaign for French chemistry energized all those associated with it, but the growing contentiousness of public life and the political disruption of the republic of science soon blocked further development of the Lavoisian chemical community. In 1782, while pursuing his heat research with Laplace, Lavoisier drafted a short general article on chemistry.[61] As if previewing the campaign that would begin in earnest in 1787, he wrote with evident optimism that recent rapid advances in chemistry "were the work of a very small number of learned men, most of whom were our countrymen, our colleagues, and our friends. Who can say," he added, "what wonderful frontiers their zeal and their success have opened for chemistry." But in 1792, in an historical note on his discovery of why metals gain weight when heated in air, he rejected the notion that the new chemistry was French chemistry and insisted instead that he alone was its author.

> This theory is not, as I have heard it said, the theory of French chemists, it is *mine*, and it is a property that I claim before my contemporaries and posterity. Others, no doubt, have raised it to new degrees of perfection, but no one, I hope, will seek to deny me credit for the theory of oxidation and combustion, the analysis and the decomposition of air by the use of metal and combustible bodies, [and] the theory of acidification.[62]

By the time Lavoisier wrote these bitter words most French chemists had accepted the chemical revolution as a *fait accompli*. The vigorous sense of community that had encouraged Lavoisier and his closest colleagues to subordinate their individual claims to discovery to a collective concern for the advancement of chemistry had been shattered, not by scientific opposition, but by the revolution's transformation of politics and the public culture that sustained it. Science in France, along with other forms of public life, was becoming a war of all against all; the community of French scientists was being overwhelmed by events.[63]

In February 1789 Lavoisier presented an ornately bound copy of his *Treatise*, embellished with the arms of Marie Antoinette, to the king and queen, an event that was duly noted in the *Gazette de France*.[64] A year later he sent two copies of the *Treatise* and a most interesting letter to his old friend, Benjamin Franklin in Philadelphia; one of the copies was for Franklin's personal use, the other was for the library of the American Philosophical Society, of which Lavoisier was a foreign member.[65] By 1790 Lavoisier had settled on a standard account of the chemical revolution, and the first part of his letter to Franklin proceeds like a well-rehearsed legal plea. But the letter also reflects something of Franklin's special status in science, in eighteenth-century culture, and in Lavoisier's affections. He and the Lavoisiers had been very close during the years Franklin represented both the ascendancy of experimental science and the triumph of American republicanism in a Paris beguiled by images of rustic virtues and natural truths. Franklin, like Lavoisier, was a distinguished scientist and public servant, but unlike Lavoisier he had won hearts as well as minds. The French thought of Franklin, who in 1790 was in the last year of his long life, as a man of mythical grandeur and avuncular wisdom, attributes of greatness never attributed to Lavoisier. The younger man had worked hard and achieved much, but he knew that Franklin's reputation had reached heights that his own would never attain.

The narrative of science in Lavoisier's letter unfolds in three sets of paired paragraphs. The first pair describes the overthrow of the phlogiston theory, the second the triumph of the experimental method, and the third the acceptance of the new chemistry. Consider first the nature and fate of the phlogiston theory.

> All treatises on chemistry published since the time of Stahl begin by stating a hypothesis; one is then obliged to observe that it can account passably well for all the phenomena of chemistry.
>   I and a great many of our contemporaries think the hypothesis introduced by Stahl, and subsequently modified, is false and that phlogiston,

in Stahl's sense of the term, does not exist. I undertook the enclosed treatise, which I am pleased to send you, primarily to spell out my thoughts on this topic.

Having dispatched phlogiston, Lavoisier turned to the experimental method that Franklin had employed so masterfully.

As you will see in the preface, I have tried to arrive at the truth by pursuing facts and by suppressing, in so far as is possible, speculation, which often misleads and deceives us. We should follow the torch of observation and experience whenever possible.

This approach, which has never been followed in chemistry, led me to organize my book in a way that is completely new, a way that makes chemistry appear quite like experimental physics. I sincerely hope you will have the time and strength needed to read the first few chapters, for I seek only your approbation and that of those learned men in Europe who are without prejudice in these matters.

Having taught chemistry how to employ the experimental method, Lavoisier was pleased to be able to report that the revolution was spreading rapidly throughout the world of science.

It appears to me that when presented this way, chemistry is much more easily grasped than it was before. Young men whose heads are not filled with systems eagerly master it, but older chemists reject it, largely because they have more difficulty grasping and understanding it than do those who have not previously studied chemistry.

French chemists are now divided between the old and the new doctrines. My allies include Morveau, Berthollet, Fourcroy, Laplace, Monge, and most of the physicists in the Academy. The learned in London and England are slowly abandoning Stahl's doctrine, but German chemists still cling to it. So you see that since you left Europe, a revolution has taken place in an important area of human knowledge. I will consider this revolution well advanced and nearly complete if you join us.

We will save the final two paragraphs of Lavoisier's letter to Franklin, which concern political developments, for a later chapter. As the paragraphs already quoted indicate, Lavoisier was clearly aware of the parallels between the revolution he had effected in science and the political revolution that was convulsing France. Yet we should recall that referring to major breakthroughs in science as "revolutions" was a rhetorical commonplace in the decades before 1789; in 1773 Lavoisier himself had looked forward to effecting a revolution in physics and chemistry.[66]

The important point is that Lavoisier completed his transformation of chemistry before the term "revolution" began to acquire its vastly expanded modern political meaning.[67] Lavoisier had, in his own view, completed a carefully delineated revolution in a particular science, not a transformation of science itself. Chemistry, having been reoriented along new methodological and theoretical lines, had been liberated from the sloppy modes of reasoning with which it had been burdened in the past. The publication of the *Treatise* brought Lavoisier's chemical revolution to a close; that task having been completed, chemists could press forward with renewed confidence and vigor. Lavoisier said as much when writing to Chaptal in 1791 to congratulate him on the publication of his new textbook.

> I take enormous pleasure in seeing that you have adopted the principles I was the first to announce. Your conversion, along with that of Mr de Morveau and a small number of chemists scattered across Europe, is all I ever sought and acceptance of these principles is greater than I ever dared hoped. Letters announcing new converts are arriving from all quarters and I see that the only people who cling to the phlogiston doctrine are those who are too old to take up new investigations or who cannot bend their imaginations to a new way of seeing things. All the young adopt the new theory, which tells me that the revolution in chemistry is over.[68]

Of course Lavoisier had no way of knowing that years later the revolution he effected in chemistry would be seen as representative of a much larger change in science that some call "the second scientific revolution."[69] Between the last years of d'Alembert's life – he died in 1783 – and the death of Laplace in 1827, French science largely shed its earlier overriding concern with ontological and mathematical foundations and emphasized instead the experimental and inductive methods of modern science. Attempts to construct comprehensive systems of natural philosophy gave way to the pursuit of positive knowledge in the paradigmatic sciences of physics, chemistry, physiology, and geology. Science, no longer thought of as simply a philosophical reflection on nature, became instead a collection of disciplines that are autono-

mous when examining their proper subject matter and that nego-
tiate their relations with one another when their aims and theories
overlap.

This revolution in the conception of science occurred over time and
not all fields were equally affected, yet the significance of the transformation is inescapable.[70] Auguste Comte, who in the late 1820s began
writing his hugely influential history and philosophy of science, was,
among other things, trying to give an account of science as he saw it
being practiced around him. We can now see that the change in the
image of science he was attempting to describe was as historically
consequential as the seventeenth-century metaphysical revolution.
This second scientific revolution was effected during the last thirty
years of the eighteenth century and the first thirty years of the nineteenth. It was a revolution that transformed science as a whole in much
the same way that Lavoisier transformed the science of chemistry.

Lavoisier's new chemistry did not bring about the second scientific
revolution single-handed. In retrospect, however, we can see that his
achievement in chemistry did show others how they could transform
subjects that had remained on the margins of natural philosophy into
"legitimate" sciences. His strategy of emphasizing exact experimentation, limited inductive generalization, the reform of language, and the
construction of a unified scientific community is still utilized by scientists who wish to create new fields of specialization. But we should also
remember that Lavoisier acted as he did not because he wished to
legislate for all of science. He chose the tactics he used because they
enabled him to consolidate his own achievement in chemistry. That
was as far as he wished to go as a scientific legislator.

This is not to say that had Lavoisier survived the political revolution,
he would not have joined Laplace and Berthollet in their attempt to
revive the Newtonian program for physics.[71] Circumstances change,
and the chaotic destructiveness of the political revolution well might
have convinced Lavoisier that stronger central authority was needed
in both politics and science. We know, from the "Reflections on
phlogiston," that he had harbored hopes of recasting his chemical
theories into deductive form, and we may assume he would have done
so had he been able to carry it off. But whatever position Lavoisier
might have adopted on this issue, we do know that when the attempt

to construct a neo-Newtonian natural philosophy was abandoned at the end of the 1820s, Lavoisier's liberal-positivist style of science suddenly seemed even more attractive. In the longer run, therefore, Lavoisier's chemical revolution was interpreted by later chemists as providing a compelling exemplar of the positivist view of science. Although it did not happen during his lifetime, eventually Lavoisier too attained mythical status in the history of science.[72]

8

Gunpowder and Agriculture

When Louis XV died of smallpox in May 1774, tradition and religious sanction required that, since his son had died in 1765, he be immediately succeeded by his twenty-year-old grandson Louis XVI. Although there was no need to consecrate the new king's claim to the throne, an elaborate and inevitably expensive coronation was held a year later in the cathedral of Rheims. Shortly after inheriting the throne, Louis XVI chose the physiocrat and former intendant A. J. M. Turgot as his Controller-General. Within two years Turgot's reform program was in disarray and he was dismissed. Turgot's brief ministry, which has traditionally been considered a turning point in French history, thus provides a convenient point of departure for narrative accounts of the revolution that erupted fifteen years later.[1]

Lavoisier's life was profoundly affected by Turgot's ministry. Like Turgot, he believed that France desperately needed economic and administrative reform, and when his assistance was requested, Lavoisier promptly joined Turgot's campaign for a top-down rationalization of the royal government. Lavoisier knew that taking on additional duties would reduce the time he could spend on his chemical research program, which was just beginning to yield striking results, yet he also realized that reformers must seize opportunities when they appear. This is why Lavoisier, at the beginning of what turned out to be the last reign of the old regime, took on the task of rationalizing the production of gunpowder in France. A few years later he began an equally ambitious private research program in agriculture.

At the very outset of his ministry Turgot urged the king to simplify and secularize the oaths administered during the coronation and to settle for a modest ceremony in Paris, where the public spectacle would reinforce devotion to the monarchy and impress foreign visitors. Vowing to extirpate heretics, as the ancient ritual required, and indulging in all the archaic rites associated with a coronation would hardly signal royal enthusiasm for the liberalizing reforms Turgot had already begun to introduce. Turgot was also worried about the cost of mounting a sumptuous royal pageant at a time when parts of Paris and the surrounding countryside were wracked by riotous protests against the high price of grain. But the king was determined to celebrate with a public spectacle of unimpeachable orthodoxy and unforgettable grandeur the event that made him the representative of all Frenchmen, an occasion one lawyer described as his "national election." The ceremony that took place was, by all accounts, as affecting as he wished it to be, but its costs, both to the treasury and in terms of the labor required of those who made the necessary preparations, did not pass unnoticed.[2]

By the time Turgot was in a position to begin eliminating restrictions on the grain trade throughout France, a policy he considered crucial to the success of his larger program, many thought it was an idea whose time had come and gone. In the 1750s and 1760s the physiocrats or, as they were also called, the economists had articulated a simple and appealing program for increasing the nation's wealth. It was founded on three assumptions: that agriculture is the sole source of national wealth, that the best way to increase agricultural production is to eliminate barriers to free trade in grain, and that a single tax on cultivated land should replace the crazy quilt of direct and indirect taxation that seriously impeded the growth of the nation's economy.[3] Restrictions on the grain trade were in fact drastically reduced in the 1760s, but the experiment ended badly.[4] Despite several good harvests, the movement of grain in response to price created local dearths. Popular reactions to profiteering, to shortages, and to rising prices led to the seizing of shipments and forced sales of grain at "fair" prices. In 1770, when Louis XV's Controller-General Terray restored calm by reinstating many of the traditional restrictions on the grain trade, the problem of food riots was laid to rest for the remainder of the reign.

In 1774, when Turgot removed the restrictions that Terray had reimposed, bad harvests magnified the dire consequences of his policy. The situation became insupportable in the spring of 1775, when stocks of flour were at their lowest, and rioting, seizures, and forced sales again became commonplace. Turgot responded with vigor, not by moderating his policy, but by protecting the merchants' right to move and price grain as the market indicated. Twenty-five thousand troops were deployed, summary trials were held, and exemplary executions were carried out. If agriculture was the sole source of national wealth and if its products had to be traded freely to make it flourish, then the government was justified, Turgot believed, in using force to suppress opposition based on customary expectations.

While undeniably reasonable, this privileging of the market, a policy that has been repeatedly invoked in modern history, could only be implemented by a strong, confident government and at the cost of considerable hardship to the poorer members of the nation. A government that was attempting to transform agriculture by eliminating traditional restraints on trade and price also had to be concerned with forestalling famine and maintaining public order. Lavoisier accepted these public responsibilities as fundamental if unstated facts of political life during the years he labored to improve both the production of gunpowder and France's agricultural productivity.[5]

In midsummer 1775, shortly after Louis XVI's coronation, Turgot, acting on a report that Lavoisier had prepared at his request, appointed Lavoisier as one of the four directors of a newly established Gunpowder Administration (*régie des poudres*), a royal office charged with managing the production of gunpowder.[6] Lavoisier was highly qualified for this position as a scientist, as a financier, and as an administrator, and he immediately added these new responsibilities to the many other activities that filled his days. By September he was deeply involved in directing the resources of the Academy of Sciences and the insights of modern chemistry to the task of improving the quality and output of nitre, an essential ingredient in gunpowder.

Lavoisier's father died in the same month he was appointed to the Gunpowder Administration. The venerable old man had been living on a small farm in Le Bourget, a town near Paris, and Lavoisier wrote that his death deprived him of a confidant who had been as dear as a friend as he had been as a father.[7] Lavoisier had inherited the farm at Le Bourget from his mother and he became increasingly interested in agriculture after his father's death, but it was several years before he found the time needed to consider seriously how agricultural productivity might be improved. His most pressing concerns in the final months of 1775, in addition to continuing his scientific research, were to organize the work of the Gunpowder Administration and to furnish the new apartment and chemistry laboratory provided for him at the Paris Arsenal. When the latter task was completed the following spring, Lavoisier and his wife moved from the house his father had given them when they were married and settled into the quarters at the Arsenal they were to occupy for the next sixteen years.

Gunpowder is made from a mixture of nitre (potassium nitrate), charcoal, and sulfur, the standard ratio being six to one to one. When ignited it burns rapidly without access to the atmosphere because, as Lavoisier pointed out in the 1770s, the nitre (KNO_3) releases oxygen.[8] The nitre used in gunpowder was extracted by solution and crystallization from saltpeter, a mineral salt that was collected, as its name suggests, from rocks and other earthy sources. Gunpowder is made by grinding the three constituents together in a mill. The grinding process was easily mechanized, but it is very hazardous. Grinding produces considerable heat and the mixture, although normally kept moist while being ground, is liable to explode at any moment. Powders of different degrees of fineness are produced for different types of weapons and other uses.

Given the importance of gunpowder in the history of colonization and warfare, European monarchs were naturally concerned that they have access to adequate supplies of saltpeter. Saltpeter in its natural state contains several different salts and can be found in a variety of locations, many of them unsavory. It crystallizes on the floors and lower surfaces of damp and fetid buildings, such as barns, dovecotes, and privies, that are exposed to the vapors arising from animal excreta and rotting vegetation. It also forms as calcium nitrate on moistened surfaces of limestone. Saltpeter occurs naturally as well in certain limestone formations and soils.[9]

Saltpeter is harvested by first breaking up the stones and soils in which it is found and then placing the pieces in large barrels so that their salts can be leached out. Potash (potassium carbonate) in the form of ashes from burned wood or seaweed is added to the solution to convert the calcium nitrate into potassium nitrate. The liquor is then boiled down and clarified in large caldrons. Common salt, being less soluble than nitre at high temperatures, crystallizes out first. At the end of the first concentration the crude nitre appears as large yellow crystals. Further cycles of solution and crystallization were needed to produce the flour-like powder that was dried into loaves and sent to mills to be made into gunpowder. The yields of nitre were slight compared to the weight of soils and stones used. In India, where the richest natural deposits were found, yields of 3 to 5 percent were realized, while the best European yields remained below 1 percent.

The manufacture of gunpowder, and the production of nitre for this purpose, was a royal monopoly in France and other European states.[10] Prior to the reign of Louis XIV the royal arsenals were in charge of gunpowder production. The military ambitions of the Sun King soon overreached the capacity of the existing system, however, and his minister Colbert decided to increase gunpowder production by leasing out the enterprise. The gunpowder "farmers," like the tax farmers, signed an exclusive lease with the government that required them to collect and refine saltpeter and sell the crude nitre to the government at a fixed price. Their agents, when engaged in this work, could enter and search farm buildings and houses and dig out whatever saltpeter they found, a privilege called the *droit de fouille*. They were also given the right to command lodgings, tools, fuel, meals, and transport, for which they paid at fixed rates that were well below market value. The saltpeter collectors, having been given the power to harass and disrupt, were notoriously willing to accept bribes in return for leaving homes and farmsteads undisturbed. In the limestone basin of the Paris region saltpeter was collected primarily from building rubble; elsewhere in France it was mostly scraped from the walls of barns and privies.

Although gunpowder farming employed many fewer agents than

tax farming, its privileges and profits made it highly unpopular. Its operations were distorted in ways that are characteristic of closely regulated and badly managed monopolies. The farm was obliged to sell crude nitre to the royal arsenals at a price that covered less than half the cost of production. When this fixed price was compared to the cost at which the government sold gunpowder to hunters, slave traders, and other consumers, it appeared that a huge profit was being made. In fact, however, the government indirectly subsidized the cost of production by providing some materials, such as potash, at artificially low prices and by rigging some markets, such as that for the crude table salt produced during refining, so that they paid artificially high prices. The gunpowder farmers therefore made handsome profits not by increasing the scale of production or lowering their costs, but by exploiting the privileges associated with their monopoly. When Lavoisier took a close look at this operation he was appalled.

The royal government might have overlooked, as governments so often do, both the economic irrationality of the way its military forces were supplied and the complaints of its citizens if the system had in fact delivered adequate quantities of good gunpowder, but such was not the case. The gunpowder farmers exploited the situation by demanding special subsidies from the government and the government, under the lash of military necessity, found itself obliged to grant them. These subsidies were used to pay for nitre purchased at high prices from the Dutch, who imported the commodity from India. It took a war, in this case the Seven Years War (1756–63), to reveal the full extent of the nation's vulnerability.

After losing her Indian colonies to Great Britain, France faced the prospect of becoming utterly dependent upon the Dutch and other foreign suppliers for nitre. This was the military and economic crisis that moved Turgot and Lavoisier to action. In a very few years the new arrangement they devised for the domestic production of gunpowder was performing its primary functions with complete success. Quite soon France was able to supply gunpowder to the North American colonists during their revolution against Great Britain. The system established by Lavoisier, when expanded, also supplied gunpowder to France's armies and navies during their far-flung revolutionary and Napoleonic campaigns.

In May 1775 the lease granted to the gunpowder farmers was revoked and their duties were transferred to the newly created Gunpowder Administration. Lavoisier had drafted the legislation that created
the new Administration and was appointed one of its first four directors. His managerial experience and his financial resources were as
important in qualifying him for this position as was his mastery of
chemistry, for the new administration retained many of the features of
a farm. The government, operating as always at the limits of its credit,
lacked the funds needed to buy out the leaseholders. Each of the directors was therefore obliged to provide the government with a loan of a
million livres upon taking office. Two of the other three directors had
held leading technical positions in the gunpowder farm and the third
had owned part of a share. Evidently they, like Lavoisier, had sufficient
personal financial credit to work for the king. In addition to annual
salaries of 2,400 livres and quarters in the Paris Arsenal, the directors
were to receive interest at 1 percent above the official rate on their loans
to the government and bonus payments when the production of nitre
and gunpowder exceeded stated minimums. All profits realized were
to remain with the government and abuse of the *droit de fouille* was to be
suppressed.[11]

Lavoisier, clearly the most vigorous and commanding of the directors, immediately launched a four-pronged attack on the problem of
providing France with ample domestic supplies of nitre. As a chemist
and a leading academician, he sought to bring the resources of science
to bear, his hope being that a better understanding of how saltpeter is
formed would prove useful to those seeking to expand its harvest. As
a naturalist he drew on his extensive personal knowledge of the mineral resources of France while searching for previously unexploited
sources of saltpeter. As an experimentalist he began a large-scale program of research designed to teach the French how to cultivate saltpeter in artificial beds, a mode of production used elsewhere in Europe.
As a manager he sought to improve the purity and yield of the nitre
already being produced by improving the procedures employed, requiring more accurate record-keeping, and instructing and supervising
the operators involved.

All this was to be done at a reduced cost to the government and with
a minimum of disruption to those who lived in areas where saltpeter

was collected. Lavoisier's strategy was characteristically comprehensive and ambitious, and it is hardly surprising that he was not completely successful on all fronts. Yet the overall situation was turned around rapidly, and in a very few years Lavoisier was able to report that the production of gunpowder had become an area of national strength rather than vulnerability. He took great and deserved pride as a scientist, manager, and citizen in this achievement. Turgot did well in matching the man to the task; Lavoisier did well in demonstrating how Turgot's program of reform could benefit the nation.

The Academy of Sciences became involved in the saltpeter campaign as the sponsor of a prize essay. In August 1775 Turgot, after consulting with Lavoisier, sent the Academy a letter asking that a committee be formed to draft and publish a notice announcing a competition that would stimulate scientific investigation of saltpeter.[12] A bit of seduction was involved, as always. Lavoisier and Turgot both knew that most members of the Academy would be reluctant to commit themselves to a practical problem, no matter how great the urgency. They would instead be looking for assurances that the proposed competition was primarily concerned with the acquisition of natural knowledge, that being their overriding interest as scientists and academicians. Turgot's letter pointedly addressed this concern; indeed to the late twentieth-century reader his letter looks more like a governmental request for proposals for scientific research projects than a royal command to take up a pressing public problem.

Turgot wrote of the importance of the issue to be addressed, the king's great interest in the problem, and the need for prompt action, but he carefully avoided specifying the type of solution sought. The prizes offered were unusually large: 4,000 livres for the winner and two honorable mentions of a thousand livres each. Turgot's letter focused on the advancement of understanding rather than the solution of technical problems of production. Contestants were to be encouraged to discover well-grounded facts capable of informing those who would be designing new processes for the production of nitre. A prize competition, it was thought, was the best way to reveal "the secret of nature by which saltpeter is formed" and the most prompt and economical way to extract it in large quantities. To ensure that equal attention would be given to experiment and theory, a stipulation that Lavoisier probably insisted on, the king was prepared to provide the members of the Academy's committee with the resources needed to examine the contestants' experimental claims. The members of the committee were strongly encouraged to work cooperatively with all those who addressed this pressing problem.

The Academy responded promptly to Turgot's request and immediately appointed a prize committee consisting of Macquer, Lavoisier, Sage, Baumé, and the chevalier d'Arcy.[13] Lavoisier believed that serving on the committee while also holding an appointment as a director of the Gunpowder Administration raised no difficulties, for, as he pointed out to those who muttered about conflicts of interest, his only interest as a director was to serve the nation. The effort to revitalize the production of gunpowder in eighteenth-century France, like the production of the atomic bomb in the twentieth century, appeared to require the centralization of authority under a field marshal capable of directing the forces of science and technology to a common purpose. Such in any case was the role Lavoisier constructed for himself, and he had little patience with those who saw what he considered service to king and nation as self-interested grasping. It was not long, however, before petty disputes arising from envy and privilege were overshadowed by the several experimental programs Lavoisier initiated and the many publications that began to pour forth from the "crash program" he directed.

The arc of Lavoisier's engagement with the saltpeter problem can be traced by plotting the series of books and memoirs he produced.[14] The series begins in 1776 with a collection of twenty-two papers culled from a wide range of learned-society transactions and other sources. These essays and reports summarized what was known about the chemistry of saltpeter and the way it was collected and processed in France, in other European nations, and in such distant countries as China, India, and America. The collection appeared as a publication of the Academy's prize committee; its purpose was to make available to those interested in competing for the Academy's prize what was known about this subject. Lavoisier, the actual editor, took the opportunity to include in the collection the memoir on the presence of air in nitric (then called nitrous) acid that he read to the Academy in April 1776.

In many European nations saltpeter was cultivated artificially in beds made up of layers of porous stony material, rotting vegetation, and animal excreta. Saltpeter will form in such a compost if it is kept moist and turned regularly, and it seemed a promising way to supplement the harvest from other sources. Although the artificial production of nitre was not much practiced in France, Lavoisier reported with evident pride that the oldest beds of this sort were located just outside Villier-Cotterets, the Lavoisier family's ancestral town.[15] An extensive experimental program on the artificial production of saltpeter was begun and a number of open sheds were erected near the Arsenal for this purpose. In 1777 the directors of the Gunpowder Administration published a small book of instructions on the cultivation of saltpeter and how best to extract nitre from the crude salt. Once again it was a work that appeared as a collective effort but actually came from Lavoisier's pen.[16]

Two of Lavoisier's four avenues of attack proved to be of practical value: the search for new sources of saltpeter in limestone formations and, by far the most important, the use of improved techniques of extraction.[17] Lavoisier soon came to appreciate that it was critically important to make better use of the saltpeter sources already being exploited and he therefore concentrated on drawing up ever more detailed instructions for extracting and refining nitre. In 1778, again under the aegis of all the directors, he published a small book designed to supplement the instructions published a year earlier. He was particularly concerned with determining the nature of the salts in solutions containing saltpeter and in the proper use of wood ashes when converting calcium salts into potassium salts. He addressed this crucial problem again in greater detail in 1779, in another small volume that was also issued by all the directors.

Despite considerable marching and counter-marching, little progress was made on the scientific front. The prize competition was supposed to close at the beginning of April 1777, but none of the thirty-eight submissions received by that deadline was deemed prize-worthy. Hope was not abandoned, however, and since no better institutional mechanism for promoting the discovery of new knowledge had been identified, the deadline was extended by five years and the first prize award was doubled to 8,000 livres.[18] These unusual steps indicate how serious the problem was thought to be, yet by the time the second deadline had passed the contest no longer had much bearing on the way the problem that occasioned it was actually resolved. The prizes were awarded and Lavoisier took on the task of editing the results.

The large volume he put together was published in 1786 by the Academy of Sciences, and while not Lavoisier's last word on saltpeter, it in fact marked the end of the campaign he had been captaining for over a decade. The first part of the volume consists of a history of the competition, runs to nearly 200 pages, and includes, among other items, abstracts of the sixty-six papers submitted. Several lengthy memoirs, including the prize-winning essay by the Trouvenel brothers, the essays by the two second-prize winners, and a few of Lavoisier's brief pieces on saltpeter, were published in the volume's more than 400 additional pages. Although victory in the gunpowder campaign was not achieved through the deployment of science, science had been enlisted into the ranks and it shared in the glory.

Two years later, in 1788, an accident involving a novel form of gunpowder provided a harsh reminder of the dangers associated with this kind of research. While investigating various chlorine compounds, Berthollet had examined a salt containing potassium, chlorine, and oxygen. He found that when heated, this oxygenated muriate of potash ($KClO_3$), as he called it, gave off pure oxygen. Gunpowder made by mixing the chlorate with carbon was much more powerful than that made with nitre.[19] When Lavoisier learned of this new material he decided to have a batch prepared at the gunpowder mill at Essonnes for testing at the Arsenal.[20]

In October Lavoisier, in the company of his wife and several others, journeyed to the country to observe the grinding of the new kind of powder. A barrier had been erected to protect the workmen turning the mill, but Le Tort, a fellow director and a man of considerable experience, was nonchalantly working directly beside the pot of moist powder while trying to speed the process along. Lavoisier later reported that he had given strict orders for everyone to stay behind the barrier, but evidently at the time the others did not share his concern. Suddenly, while Lavoisier and most of the party were fortuitously at some distance, the mill exploded, mortally wounding Le Tort and a young woman standing nearby. It is usually thought that scientists risk only their theories; clearly Lavoisier's commitment to public service exposed him to immediate hazards to his health as well.[21]

By the end of 1788 the Gunpowder Administration was manufacturing all the gunpowder France needed. When the Administration was established in July 1775, the gunpowder farm was producing less than half of the 3,600,000 pounds of powder required each year. The situation was turned around in very short order. In 1776 and 1777 France was able to supply the American colonies with more powder than the gunpowder farm had manufactured in an entire year. Quality improved as well, with French powder becoming the best in Europe, while the price of gunpowder declined rapidly. Clearly the Gunpowder Administration was performing its job in an exemplary manner and was saving the nation huge sums of money. France, her allies, and the cause of reform were being well served. "One can truly say," Lavoisier wrote with pride in April 1789, "that North America owes its independence to French gunpowder."[22]

If Lavoisier had reason to be proud in 1789, he had reason to be anxious in 1776, when Turgot was driven from office. Placing the production of gunpowder under the direct administrative control of the royal government was just the sort of reform that was making French advocates of liberty, however varied their political rationales, exceedingly anxious. Lavoisier shared many of their reservations, for while he had no doubts about his own ability and rectitude, he believed that in general it was better to maximize liberty rather than central authority. But his vision of the public good placed a high value on informed and efficient administration, and he was not much concerned with the abstract claims of what would later be called ideology. He believed that the proper balance between central authority and individual liberty could only be determined on a case-by-case basis.

At the beginning of Louis XVI's reign Lavoisier had taken the risk of publicly allying himself with Turgot and his program of reform. When Turgot's enemies hounded him from office in May 1776, Lavoisier feared that the recently conceived Gunpowder Administration might be aborted as well. He responded to this threat not by publishing a general treatise on the need for reform, but rather by drafting a brief, sharply focused argument on why the production of gunpowder ought to be centralized. Although he decided not to publish this essay, the way he constructed his case is revealing.[23]

Lavoisier may have planned to publish this essay anonymously, even though the identity of the author would have been apparent to all. In the first paragraph he adopted a rhetorically useful, although factually misleading, stance of uninvolved innocence. He is writing, he said, as "a citizen who insists he had nothing to do with this revolution [i.e. the creation of the Gunpowder Administration] and who knew nothing of the matter before the minister had ordered it and brought it to completion." He is, he insisted, "a person who is known to have sacrificed his own interests to state service and the love of truth." Knowing the author's identity, we are tempted to consider these sentiments craven and self-serving, but later on in the essay Lavoisier boldly praised the recently dismissed Turgot for carrying through in France "the same revolution in the production of saltpeter that occurred thirty years ago in Sweden, in Prussia, and subsequently in other states."[24] Lavoisier evidently sought to put some distance between himself and Turgot not to protect his personal interests, but so that he could more effectively defend his patron's policies.

The production and sale of gunpowder, like the processing and retailing of tobacco, but unlike other forms of commerce such as the grain trade, was a government monopoly. This did not mean that the government was unaccountable, however, for as Lavoisier acknowledged in a strikingly liberal aside, all constraints on freedom must be fully justified. "Every exclusive privilege is without question a violation of natural order. It is a blow struck against the natural liberty of the citizenry and such interventions can only be supported to the extent that they work to the advantage of the government."[25]

Lavoisier's task, as he saw it, was to provide convincing evidence that the new Gunpowder Administration met this test. He demonstrated, for instance, that the cost of powder sold to the public was not too high, rather the costs of producing powder under the farm had been grossly understated. He urged that France, like other nations, promote the artificial cultivation of saltpeter so that the inconveniences of the *fouille* could be eliminated. He asserted that France's gunpowder needs could best be met not by relying on "a company of entrepreneurs," but by putting powder production under the control of well-instructed administrators working directly for the king. The arguments are straightforward. What is remarkable is that as early as 1776 Lavoisier was prepared to engage the issue of liberty versus central authority and make his case in public. In the event, the Gunpowder Administration survived and Lavoisier was given the time he needed to demonstrate that his defense had substantive merit and was not merely rhetorical or self-serving.

Lavoisier was a shrewd, determined, methodical strategist. Until the end of Louis XV's reign he divided his time between two fields of endeavor. As a scientist he explored a number of areas of research, almost all of which he addressed under the aegis of the Academy of Sciences, and he succeeded in becoming an eminent academician and chemist. As a member of the Company of Tax Farmers he devoted a great deal of time and energy to administrative problems and, in the course of making a name for himself as a manager and financier, he increased his personal wealth considerably. At the beginning of Turgot's ministry Lavoisier agreed to take on new and more overtly political duties. In accepting responsibility for the production of gunpowder, he was doing much more than offering to help solve a particular technical and administrative problem. He was also publicly declaring his commitment to Turgot's program of reform, and once having done so there was no going back.

Turgot's dismissal in 1776 therefore posed two distinct threats to this highly political extension of Lavoisier's administrative career. His immediate concern was to ensure that the reform of gunpowder production, on which he had just embarked, should not be frustrated by political opposition. Once he had countered that threat, he began searching for other ways to press forward with the kinds of reform that he and his fellow Turgotists still believed were needed. He knew this campaign would have to be carried out in a political culture riven with suspicion and hostility. The major long-term challenge he faced following Turgot's fall was to relegitimate his royalist, centralizing, and science-based approach to public policy and public administration. The reform of agriculture, Lavoisier decided, offered a particularly promising line of attack.

Those interested in improving French agriculture had traditionally conceptualized the problem as either technical or political; Lavoisier's goal was to bring about an intersection of these two views. According to the historian of French agriculture in the eighteenth century, the reforming efforts of the agronomists and physiocrats were rooted in separate traditions that seldom overlapped.[26] The agronomists were mostly farmers, prosperous rural gentry who had an interest in improving agricultural productivity. Being local men of enlightenment, they were aware of the innovative farming techniques that were being introduced in England. They were also in a position to undertake the long cycles of experimentation required to determine whether new modes of farming were in fact capable of increasing yields. They described their efforts in letters and reports that circulated widely and their findings were periodically gathered up and published in encyclopedias and handbooks. This literature commanded attention not because it was scientific or useful; indeed there is little reason to think it had any practical consequences. Its chief significance was political; it embodied the discourse on agriculture initiated and sustained by those who dominated rural France.

The physiocrats were men of theory. They sought to improve agriculture by altering the circumstances in which farming took place rather than by addressing directly its techniques of production. Lavoisier shared the urban, theoretic, and public policy outlook of the physiocrats, but he knew that some of their generalizations were insupportably doctrinaire.[27] Being a chemist, he was also more intensely interested in the theoretical and technical aspects of agronomy than were his fellow economists. Agriculture was, of course, the foremost source of national wealth, but Lavoisier, like Colbert, realized that commerce and industry contribute to the nation's wealth as well. Lavoisier was in favor of liberating the grain trade and rationalizing taxation, but he also realized that farming practices could be improved only through direct intervention at the point of production. Physiocracy's greatest weakness, however, was its political isolation. Following Turgot's dismissal Lavoisier evidently realized that programs of reform that originated in Paris were not likely to succeed unless they incorporated the experiences and modes of understanding of those whose lives they were designed to affect. The agricultural remedies proposed by Quesnay, Dupont, and Turgot were systematic and rational, but in the eyes of those who farmed they were also often viewed as excessively abstract. Lavoisier's program for agricultural research was designed to bridge this gulf.

Lavoisier no doubt discussed these issues with Pierre Samuel Dupont de Nemours during the months following Turgot's fall. Turgot's political failure had been far more disastrous for Dupont than for Lavoisier, for they had long been political allies. In 1764 Turgot was serving as the royal intendant in the Limousin when he read Dupont's tract on the exportation and importation of grain; when he next got to Paris he made a point of getting acquainted with the young man who had already developed quite a reputation as the spokesman of physiocracy.[28] The two men remained close thereafter.

When Dupont's second son was born in 1771, Turgot agreed to be his godfather and suggested naming him Eleuthère Irénée, from the Greek for liberty and peace. This was the son who, with Lavoisier's endorsement, learned how to grind gunpowder at Essonnes and later suggested naming the Dupont powder works he established in Delaware the Lavoisier Mill.[29] When Turgot became Louis XVI's Controller-General, P. S. Dupont joined him as a secretary. When Turgot was dismissed, Dupont was exiled to his country estate, this being a way to give the new ministry a chance to establish itself. For two years he devoted himself to practical farming and cultivating new connections in the capital. In 1778 Dupont was recalled to Paris to serve in the finance ministry. Soon thereafter he successfully promoted the idea that the office of taxation should concern itself with agricultural reform. It appears likely that by that time Lavoisier and Dupont were working together to formulate a new policy for reforming French agriculture.

In 1778, the year in which he was also promoted to the rank of pensioner in the Academy of Sciences, Lavoisier purchased a large estate, Fréchines, that lies between Blois and Vendôme in the Loire valley.[30] It is tempting to see this acquisition as nothing more than another instance of a successful bourgeois investing in rural property, but Lavoisier clearly had something more in mind as well. He was not a man who sought leisure, nor did he plan to divide his time, as Buffon did, between being lord of a manor in the country and master of a scientific coterie in the city. Lavoisier's new farm was substantial but run-down. He bought it, he later reported, because he wished to begin an extensive program of research, his goal being to determine how best to improve the productivity of a farm having poor land and a small number of animals.

This was not a casual or opportunistic undertaking. The plan of research Lavoisier drew up was detailed and exact; the monitoring of methods and outputs was carried on for nine years before the results were made public.[31] In the end, Lavoisier reported, the rates of return on investment in a well-managed farm were so much less than those available from financial investment that fiscal reform had to precede agricultural reform. But well before reaching this dismal conclusion he had shown that he was addressing agricultural reform not as an urban theorist, but as someone who had extensive first-hand experience. His proposal that agriculture be improved through the union of science and administration was not an exhortation deduced from a sweeping theory of political economy; it was a substantive policy recommendation firmly grounded on experimental facts.

If Lavoisier's interest in agriculture was, for him, a new departure, it also represented yet another in a series of activities that extended his reach well beyond Paris. Early in his scientific career, while still trying to decide how to focus his interests, he had enthusiastically joined Guettard on long mineralogical excursions; in the 1780s Lavoisier was still trying to bring their mapping project to completion.[32] The tours Lavoisier had taken for the Company of Tax Farmers and the trips he made while trying to locate new sources of saltpeter gave him additional opportunities to catalog the resources and observe the ways of life of rural France. Thus when pursuing his interest in agriculture, he could draw on a considerable fund of knowledge about the provinces of France, all of which had been collected under the spur of scientific curiosity and a commitment to the public good.

When Lavoisier and Dupont were ready to carry their agricultural campaign into the political arena, they chose as their vehicle a newly constituted Committee on Agriculture.[33] This Committee was a creation of the royal finance ministry and addressed the same set of problems that had long occupied an older and still active Society of Agriculture. The Society had been founded by the minister Bertin in 1761 and in its own unobtrusive way did what it could to promote and propagandize for agricultural improvement. Lavoisier joined the Society in 1783; Dupont was also one of the fifty-eight resident members enrolled in it by 1788. But Dupont and Lavoisier were not prepared to settle for persuasion and fellowship, and in June 1785 they convened the first meeting of their new Committee on Agriculture.

This Committee consisted of a small group of activists eager to take charge of the issues with which the Society continued to be concerned.[34] The Committee's membership grew from the original six members to twelve during its two and a quarter years of activity; its meetings were held at the same time as those of the Society. The finance ministry instructed the Committee to examine proposals for agricultural improvement that had been submitted to the government, but it soon ranged far beyond this standard charge and began initiating projects of its own. Lavoisier served as secretary, a position that enabled him to energize the Committee and direct its attention to areas of concern that appeared urgent and amenable to governmental intervention.[35]

Dupont insisted that the Committee on Agriculture was not gratuitously duplicating the work of the Society. The Society, like other specialized local academies, was concerned solely with its immediate region; the Committee, being a governmental body, was concerned with the nation as a whole. The Society promoted the collection and dissemination of agricultural knowledge; the Committee was concerned with the administration of agriculture as well as the increase of knowledge. Diffusing agricultural enlightenment throughout the countryside was, the Committee realized, a daunting task and the members, worried by the cost of printing, considered the possibility of conveying instruction through the clergy at the parish level.[36] But in the end the Committee's proposals were doomed by lack of support from higher levels of government, a neglect rooted in the government's increasing impoverishment rather than the infeasibility of the programs proposed.

The Committee soon realized that identifying needs and opportunities and formulating rational policy proposals would not be enough to command resources. Its perhaps unavoidable response was to devote more and more time to the unpromising strategy of courting powerful ministers. During the latter months of 1787 the Committee met less frequently and its proposals became less confident; by the end of the year it had lapsed into inactivity. It had correctly analyzed the administrative constraints that locked French agriculture into cycles of impoverishment and had proposed a series of remedies that deserved to be tried, but through no fault of its own it had the misfortune to be operating in a era of declining political power. Lavoisier and Dupont were not wrong to urge the government to seize the initiative and reinvigorate French agriculture. Yet because their program came to nothing, one is tempted to fault them politically for advocating excessive centralization and to suggest that a more decentralized regional strategy might have enjoyed greater success. They would have vigorously disputed such a conclusion. Unhappily for all concerned, many of those who vied for power at the end of the old regime were all too willing to see every royal initiative as evidence of tyrannous intent.

The demise of the Committee on Agriculture left the government without a body to examine proposals for agricultural improvement. In 1788 the Society of Agriculture, having survived the Committee's challenge, was designated a royal society and assigned these duties. Lavoisier continued to work with this group but did not play a leading role in its affairs. He had by then devoted considerable personal time and funds to reforming French agriculture and he had put his reputation as a scientist and administrator at risk as well. Although he did not know it at the time, the serious game of politics in which he and Dupont had been involved would soon be brought to a sudden close and they and their compatriots would be swept up in a revolution that would quickly lay bare the extent of France's political collapse. The Turgotists had played their cards with skill and dedication, but they had not been able to dominate the game of politics as it was played in the old regime.

French political life was transformed during the early years of the Revolution and many prominent public figures, including Lavoisier, welcomed the new spirit loosed by the upheaval. Those who had been frustrated by the ineffectiveness of royal governance took heart at the removal of traditional impediments to rational administration. Although Edmund Burke's prescient *Reflections* would later compel assent, at the outset it was not unreasonable to think that the forces released by the Revolution could be harnessed to the task of completing the Turgotist program of reform. Dupont and Lavoisier were not alone in applauding the recovery of will in public life that accompanied the triumph of the third estate during the first years of the revolution, and they continued to look for opportunities to implement the policies they had formulated following Turgot's dismissal.

In 1791 the National Assembly, in which Dupont sat as a representative, provided Lavoisier with one such opportunity. He was asked to present a report on the aggregate wealth produced annually in France, a topic he had been investigating for nearly two decades. Certain comments Lavoisier made when introducing his report reveal with astonishing frankness his views on the progress of science and its application to public policy. These were evidently deeply held if seldom stated beliefs, and they informed the administrative proposals he championed in both the final decades of the old regime and the early years of the Revolution. If today we find his views theoretically inadequate and politically technocratic, we should at least acknowledge that to him they appeared to be novel, true, and useful. We should also acknowledge with some humility that over 200 years later we are still wrestling, all too often with little success, with the same kinds of political issues that Lavoisier believed could be transformed, by the application of science, into manageable administrative problems.

Lavoisier began calculating what can loosely be called the gross national product of France in 1784.[37] He became interested in this problem after hearing a report Dupont had prepared on the subject for the Committee on Agriculture, and having once looked into the question, Lavoisier never let go of it. His interest was far from abstract. The economists were intensely concerned with the reform of taxation, their guiding principle being that only net income should be taxed. For those who believed that agricultural production was the sole source of wealth, this meant that the nation should impose a single tax, one that would capture for public purposes an appropriate portion of the net agricultural income. Lavoisier made this assumption explicit by defining the net revenue as the part of the gross national product that should be "divided between the public treasury and the proprietors."[38] The task Lavoisier set himself was to provide an accurate account of the annual net income of the nation so that a suitable rate of taxation could be determined.

The task could be completed, Lavoisier believed, because, given enough information, economics can be reduced to accounting. His faith in the power of positive facts was evidently equal to that of modern advocates of input-output models. "Please allow me to observe," he told the National Assembly.

> that I have given you a few examples of the kind of analysis and calculation that can provide a foundation for all of political economy. This science, like nearly all others, began with discussions and metaphysical reasoning. Theoretically it is quite well-developed, but the practical science is in its infancy. The public administrator (*homme d'État*) at all times lacks the facts he needs to support his proposals.[39]

The point is made again even more emphatically at the end of the introduction: when the facts are available, political economy is no longer subject to dispute and hence ceases to be a science.

> A work of this sort presents in a few pages the entire science of political economy, or to put the point a different way, this science ceases to be a science. Its results are so clear and evident and it is so easy to answer the different questions one can put to it that one can no longer hold different opinions on these issues.[40]

Lavoisier realized, of course, that his tabulation of the gross national product was not perfect, but he believed the remaining errors of omission and commission did not invalidate his main claim. "Time," he announced confidently, "furnishes the means needed to rectify such errors."

Lavoisier was clearly deploying this view of science for rhetorical purposes, but his rhetorical strategy conforms so closely to his investigative research programs that one is justified in concluding that he believed his rhetorical account of science was in fact an accurate depiction of the enterprise. Science begins with speculation and ends in facts; it begins with contention and ends with consensus. The accumulation of indisputable facts, not logic or mathematics, carries science across the divide that separates opinion from truth. Mature science, when coupled with political authority, brings peace; to help science reach maturity the government should support agencies that collect and make available relevant bodies of fact.

Lavoisier believed he had shown the way in political economy, but no individual could complete the task alone. The government should create a bureau charged with collecting information about the national economy. By enlisting the assistance of other officials a bureau of this sort could "reach with ease to the outermost branches of the tree of politics, even to the municipalities. By utilizing a patriotic correspondence undertaken for this purpose, it could obtain all sorts of information and undertake all sorts of tasks."[41] As facts transform science into positive knowledge, so can science transform politics into administration.

Lavoisier urged the National Assembly to act promptly and decisively. "To set up a great institution of this sort, such as exists in no other nation and can only exist in France, the National Assembly only has to express its desire and exercise its will. The kingdom is currently organized in a way that will facilitate such research."[42] An agency of this kind would provide the nation's political leaders with invaluable information they could use to model the economic and social consequences of political decisions.

> General accounts of this sort, which could be extended to include studies of the population and the balance of trade, would provide a reliable thermometer of public prosperity. Each legislature would immediately be able to see for the nation as a whole the good and bad consequences of the actions of earlier legislatures.[43]

Lavoisier's optimism, his confidence that political good can be wrung out of political disruption, was unshaken by the early events of the Revolution. The scientist motivated by a passion for truth and the administrator motivated by a passion for the public good appealed with unusual directness to the National Assembly, the newly constituted voice of the people. In the heat of the historic moment, beliefs and hopes that had earlier been carefully masked were publicly displayed. For this the biographer and historian is most grateful.

It should also be noted, in response to those who would reduce all motivation to selfish interest, that Lavoisier's plea for a rational system of taxation was delivered less than a week before the National Assembly voted to terminate the lease of the Company of Tax Farmers.[44] He cannot have been unaware that the Tax Farm's days were numbered. Had Lavoisier been determined to protect the partnership to which he had given so much and from which he had received so much in return, he would not have presented such a forceful case for physiocratic tax reform. It seems likely, however, that as early as the mid-1770s he had accepted that the indirect taxes collected by the Tax Farm were regressive and should be eliminated.[45] But of course he also knew the older tax system could not be abandoned until a more rational policy was in place. Like a successful capitalist professing to be a socialist in principle, Lavoisier kept working to make tax farming as efficient as possible while looking forward to its elimination. Contradictions are more easily eliminated from arguments than from the affairs of everyday life, and in times of revolution consistency is more honored in principle than in practice. Lavoisier was prepared to accept the elimination of the Tax Farm; what he rightly feared was the anger of the people's representatives if the government should suddenly find itself without funds.

9

Mesmerism and Public Opinion

Picture this. It is early summer in 1784. Lavoisier, along with other members of a committee appointed to examine the medical claims of Anton Mesmer, is sitting around a low tub in which, it is said, the substance that causes animal magnetism is concentrated. Other members of the committee are its chairman, the eminent astronomer Jean Sylvain Bailly, Dr Guillotin, whose invention would end the lives of both Bailly and Lavoisier, the chemist d'Arcet, the physicist Le Roy, and several other physicians and members of the Academy of Sciences. Benjamin Franklin, the Nestor of electricity and the beloved symbol of American independence, is also on the committee. Although living in Paris, Franklin is too frail to join his colleagues in the mesmerist's salon; he will be magnetized at home.

The committee members have gathered for one of their weekly two-and-a-half-hour sessions, their purpose being to see if they can experience at first hand the power that the mesmerists claim enables them to bring about medical crises and effect cures. For more than three months they will meet for this purpose at the clinic of Charles Deslon, a physician and Mesmer's most prominent French disciple. Anton Mesmer is not directly involved in the investigation, both because he has refused to cooperate and because, being Austrian, he is not a subject of the King of France. It is Mesmer, however, who by six years of notorious advocacy and practice has forced the government to act.[1]

The tub is about four and a half feet in diameter and is covered with a top pierced by several holes. It holds bottles of magnetized water arranged in a circle and surrounded by a bath of damp sand containing bits of iron filings and glass. Several iron rods on flexible mountings protrude through the holes in the top. Those who are to be magnetized sit around the tub with their knees touching. The flow of the magnetic substance is enhanced by a cable joining them one to another and by having everyone grasp his neighbor's left thumb between the thumb and index finger of his right hand. Once the seance has begun, the flow of magnetism is occasionally strengthened by passing one of the tub's rods over the parts of the body being treated. The attending mesmerist also intervenes by waving his wand over individuals and directly manipulating the spines and affected parts of the participants. The agent that brings about crises and cures is a subtle magnetic fluid that pervades the cosmos. To concentrate its effects, the room is closed, the curtains drawn. As Lavoisier no doubt noted, the room's atmosphere is monitored with a thermometer, a hygrometer, and a barometer. To help bring the bodies of the participants into harmony with the cosmos, musical selections are played throughout the session.

The image of Lavoisier and his colleagues submitting to hours of such unavailing mumbo-jumbo is both captivating and puzzling. These were, after all, responsible men of action, not foppish gallants having nothing better to do with their time than engage in fashionable amusements. Nor was this a serious experiment, for the outcome was hardly a surprise: the members of the committee predictably experienced nothing that in any way confirmed the existence of the supposed animal magnetism. Why then did so many men of such distinction devote so much time to an investigation so lacking in genuine scientific interest? Most scientists allocate their time in proportion to the importance they attach to the various activities in which they are involved. Late in the twentieth century successful scientists understandably seek to avoid being drawn into controversies over scientific creationism, claims that new forms of water have been discovered, or experimental reports of cold fusion. Scientists doggedly pursue their own agenda; their attention swerves only when disturbed by a powerful external force. In the case of mesmerism, that external force was a serious concern with maintaining the authority of official science in the administrative affairs of the nation.

Mesmer's flamboyant medicine shows could have been tolerated had his disciples not decided to turn them into a national enterprise. Although Mesmer frequently behaved outrageously in his pursuit of wealth and approbation, he was, so long as he acted individually, just one among a number of charlatans proclaiming they had unique powers and attracting followers among the enthusiastic and the credulous. During Mesmer's early years in Paris the Academy of Sciences knew of his presence but ignored him; the Faculty of Medicine limited itself to ensuring that mesmerism remained on the margins of medicine. But when two of Mesmer's disciples, Nicolas Bergasse and Guillaume Kornmann, formed a joint-stock company called the Society of Universal Harmony and began promoting mesmerism throughout France in what today would be called a franchise arrangement, the authorities felt compelled to respond.

The enterprising mesmerists modeled their new society on the existing network of Masonic lodges and quickly succeeded in establishing a large number of provincial affiliates. The royal officials who watched all this were unimpressed by the mesmerists' professions of devotion to the public good, for, as they well knew, France had a long history of ferocious political opposition rooted in religious and utopian sects.[2] The politics of culture and the authority of the established learned corporations were matters of governmental concern in France; the determination of truth was not left to entrepreneurs or the marketplace of ideas. For several years prior to 1784 the Baron de Breteuil, the royal minister responsible for the Paris region, and the astronomer Bailly, who was also a member of the Académie Française, had been working together on issues of public health. When they decided the mesmerists' challenge had to be addressed, they were determined, being good generals, to bring overwhelming force to the engagement.[3]

The confrontation over mesmerism is notorious for several reasons, perhaps the most important being that it provided an early opportunity for several cultural radicals, most notably Jacques Pierre Brissot, to press their campaign against the scientific establishment. Yet in 1784 mesmerism appeared to be merely one more challenge to a government going about its business of maintaining public order. Both sides realized that the issue would be joined on the field of public opinion. Such contests could be enormously consequential in Bourbon France, where rank, ritual, and reputation were powerful instruments of legitimacy. In a society so attuned to gossip and ridicule, engagements of this sort were routinely conducted with considerable cunning and indirection. The institutional and rhetorical resources deployed varied according to the subject in dispute. In the continuing struggle between the monarchy and the *parlements*, for instance, the terms of contention were borrowed from the hoary discourse of constitutional law. In the dispute over mesmerism they were appropriated from the discourse of experimental science.

Given the chronological proximity of the mesmerist dispute and the fateful year 1789, it is tempting to see the earlier event as an anticipation of the Revolution that followed so closely on its heels. The temptation should be resisted, however, for the debate over mesmerism cannot be accurately understood if it is perceived as a precursor of a political collapse that none of those who participated in it expected. It is equally important that the contest over mesmerism should not be interpreted as at bottom a conflict between an assertive group of independent practitioners and a corps of professionals seeking total control over such important social activities as the practice of medicine or the determination of what constitutes scientific knowledge. The scientists and doctors Mesmer challenged had no desire to mount a root-and-branch purge of the sort that had traumatized the nation during the long years of religious persecution. Thoroughgoing professional domination of practice may be possible in modern nation-states, but it was simply unthinkable in a traditional kingdom as large and heterogeneous as eighteenth-century France.

Those who held positions of authority in the Paris Faculty of Medicine and the Royal Academy of Sciences responded to the mesmerist challenge because they felt compelled to defend their claims to competence as men of learning and their authority as members of established corporations. That authority was both cognitive and political, as was made clear in a rather overdrawn but nonetheless revealing passage in the Mesmer committee's final report.

> The government cannot remain indifferent to questions that affect the lives and health of its citizens. According to the system of Mesmer and his disciples, everyone can, simply through the control of magnetism, cure others. If this is so, the entire science of medicine becomes useless. Its schools should be closed, its system of instruction changed, the institutions that serve as repositories of medical knowledge destroyed, and everyone should study magnetism. On an issue of such importance, the government ought to guard against too facile acceptance and absolute belief.[4]

An optimal response to mesmerism, from the government's point of view, would inflict disabling damage on its claims without imposing great costs on the respondents, and Bailly, Lavoisier, and their allies soon devised a strategy that did the job perfectly.[5] Since it was operating from a position of great formal authority, the committee had to be especially careful not to appear haughty, arbitrary, and self-serving. Mesmer had won an enthusiastic and devoted following among the fashionable public in Paris and it was that public that would ultimately decide whether or not his indictment of science was credible. Lavoisier, Bailly, and the other members of the committee spent hours at the mesmerist tub so that it could not be said they had dismissed the testimony of others without direct examination. Their highly visible attempt to experience the power of animal magnetism shifted the debate from a confrontation pitting those who had been magnetized against those who had not to a dispute over which witnesses provided the more reliable testimony about a common experience. Their strategy was to destroy the mesmerists' claims from within and in the full glare of public scrutiny, rather than to translate them into technical jargon and dismiss them from on high.[6]

Once the battle lines had been drawn, reputable scientists quickly closed ranks. The prominent veterinarian and medical reformer, Félix Vicq d'Azyr, had opposed mesmerism from the outset, but the chemist Berthollet had been an early subscriber to the Society of Universal Harmony. Later, however, when the opposition between mesmerism and science forced him to make a choice, Berthollet stormed out of a mesmerist salon and loudly denounced the claim that they effected cures by controlling a physical agent.[7] Condorcet also drafted an antimesmerist essay that conformed precisely to the established order of battle.[8] "When examining extraordinary claims," Condorcet wrote, "only those who are competent judges should be trusted." Competence, he made clear, is determined by one's standing in the republic of science, not simply by making assertions about how nature operates. While the claim was fully generalized, his specific target was clearly the well-known critic of the Academy of Sciences, Jean Paul Marat.

> When I speak of a scientist or a physician, I do not mean someone who has written books on physics or who has received a medical degree from some university. I mean a man who, well before the issue of mesmerism arose, enjoyed in France, and in Europe as well, a well-established reputation. This is the sort of witness I need if I am to believe an extraordinary claim in science or medicine.

Condorcet's little essay, though never published, is a concise catalog of the committee's arguments against mesmerism. "Enthusiasm, which longs to believe, is often mistaken," he pontificated. He also pointed out that the fad for mesmerism resembled other episodes of possession, including the famous case of the nuns at Loudun. The mesmerists' obtuseness in the face of physical and medical facts reminded him of an earlier century, in which it was said that no physician over forty would accept William Harvey's demonstration that the blood circulates. Mesmerists' arguments for the existence of the magnetic fluid do them no credit, he insisted, for while they assert there are physical interactions, they make no effort to measure or calculate them. And in response to mesmerists' objections to academic disdain, Condorcet recounted a simple tale designed to remind them of their proper place in the social hierarchy of knowledge.

> Since Mr Mesmer is unhappy with the academies, we will take the liberty of telling a little story. A man who had figured out how to square the circle complained that no one would examine his discovery. "You must understand," an academician told him, "that such examinations have repeatedly been a waste of time." "That's fine for the others," the circle squarer said, "just examine mine, which is the only good one."

Bailly and Lavoisier decided to focus their examination solely on the mesmerists' physical claims while ignoring completely all medical evidence.[9] The mesmerists objected, of course, for they considered their scientific and medical doctrines part of an integral whole: the cures they effected provided compelling evidence that the magnetic fluid is a pervasive and powerful natural agent; the theoretical explanations they offered rendered the medical effects intelligible. But their adversaries could not be compelled to examine all the mesmerists' claims, and they carefully defined what they considered to be the core issue in a way that worked to their advantage. The strategy of response Bailly and Lavoisier employed actually involved two steps. The first was to characterize the mesmerist challenge as a public-health issue, the second was to limit the challenge to a claim concerning physical causation.[10] By characterizing the issue as a medical problem the committee justified governmental scrutiny; by treating it as a problem in physics they were able to dismiss as irrelevant the mesmerists' most dramatic evidence. The mesmerists were thus forced to defend their weakest position; having been outmaneuvered, they were left sputtering.

While the dispute over mesmerism was overtly about physical causes and medical effects, at a deeper and less obvious level it brought into public confrontation two incompatible images of science. The historical importance of this subplot can hardly be exaggerated and we shall return to it at the end of this chapter. The mesmerists, with good reason, considered themselves true disciples of the Enlightenment. Brissot, like his friend Marat, and Mesmer, to a more limited extent, placed their trust in reason as they moved from the study of effects to the identification of causes. Like the *philosophes* whose ranks they longed to enter, they sought to add their voices to the discourse on natural philosophy. They therefore began with the assumption that the substances and powers that lie behind the perceptible world are entirely natural.

When constructing specific theoretical explanations, these would-be *philosophes* took as their models the philosophical heroes of their age: Descartes, Newton, and Leibniz; Voltaire, Diderot, and Buffon. How then could they be faulted? Where had they gone wrong? Since they employed canonical methods, their conclusions must be correct. Science is science and one need not be a genius or a grandee to discover novel truths. Having devoted themselves to the study of nature and having published reports of their findings, they resented not being granted the recognition and rewards they considered their due. They believed they were being victimized rather than celebrated not because their science was inadequate, but because those who controlled the leading scientific institutions were using their authority to protect their own interests. This was the charge that lay buried just below the surface of the polite debate over mesmerism and it was this charge that the committee set out to refute.

Science, the committee insisted, is not as simple as many of its devotees in the republic of letters assume it to be. One does not become a scientist simply by looking for natural rather than providential causes, nor is scientific reasoning simply a matter of being reasonable. Scientific inference and scientific verification require special skills. Science is difficult and fraught with pitfalls, for nature does not reveal her secrets readily. Few laymen realize how little we know about how nature operates. In insisting on the difficulties and hazards of science, the leading scientists of the 1780s were openly rejecting the facile naturalism of the high Enlightenment. Instead of populating the cosmos with still more subtle matters and active powers, they were stressing the need for precision, systematic experimentation, and quantification; instead of constructing yet more elaborate mathematical models, they were pointing out how difficult it is to determine the fit between mathematical abstractions and physical reality.

This shift in standards of demonstration was not abrupt, but rather a matter of degree and emphasis. The best of the earlier scientists had of course been intensely concerned with the difficulties of inference and the need for experimental confirmation. The men of the 1780s, although committed to grounding their science on positive facts, also knew they could not dispense entirely with metaphysical assumptions. Yet the shift from generalizing speculation to specifying exactitude created a clear fault line in eighteenth-century science, one that separated the confident and comprehensive materialism of the middle decades from the methodologically more restrained sciences of its final quarter. The committee took this difference as its bench mark. Speculative naturalists who made unverifiable claims were to be excluded from science, however great their popularity. Those who adhered to rigorous standards of scientific method while making significant contributions to the advancement of knowledge were to be treated with the respect due to authoritative spokesmen.

Bailly and Lavoisier were natural allies in this effort, for Lacaille had taught them both the importance of observation and logical rigor. Bailly, who was a skillful observer and calculator, entered the Academy of Sciences as Lacaille's protégé in 1763, and he soon distinguished himself with a series of papers on Jupiter's satellites. In the 1770s he wrote a four-volume history of astronomy that brought him literary success and additional academic honors. By the time he was appointed chairman of the committee on mesmerism he was a highly regarded senior statesman of French science and a man with whom Lavoisier could work easily.

As the Mesmer affair was drawing to a close, Lavoisier gathered together several of his papers, his intention being to publish them. In a brief essay included in his collection he summarized the committee's strategy. "The committee's intent," he wrote, "is to trace out the chain of reasoning so as to determine where it is broken and to focus on facts rather than reason." The mesmerists claim they manipulate a magnetic fluid, but "good logic does not allow the introduction of new principles to explain facts that can be explained by other principles that are already known."[11] The practitioners of mesmerism use two means to achieve one effect: they touch their patients and they claim to direct the magnetic fluid into their bodies; the effect achieved is that some patients experience convulsive fits. "But it is known that imagination alone, pushed or brought to a certain point, is capable of producing these effects."[12] The mesmerists err in thinking these effects provide evidence for the action of a specific physical agent.

The crux of Lavoisier's argument is that touch alone, when administered to suggestible individuals in the emotionally heightened setting of a mesmerist seance, provides a sufficient explanation of the effects observed. But before stating this conclusion he distinguished between the methods of inference employed by sophisticated scientists and the kinds of association exploited by those who know just enough science to fool the public.

> The art of drawing conclusions from experiments and observations consists of evaluating the probabilities and estimating whether they are large enough or numerous enough to constitute a proof. This type of calculation is more complicated and more difficult than is generally appreciated; it requires great knowledge and is beyond the capacities of most men.

The common misunderstanding of this sort of calculation is fundamental to the success of charlatans, sorcerers, and alchemists, as it was in an earlier age to the success of magicians, enchanters, and all those who abuse or attempt to abuse the credulity of the public.

The evaluation of probabilities is especially difficult in medicine . . . When one administers remedies, it is incredibly hard to distinguish what to attribute to nature and what to the medicine.[13]

Lavoisier was right, of course. What is historically significant, however, is that in this public dispute he employed a strategy that directly confronts the common, if groundless, belief that any fair-minded educated person can make valid inferences from effects to causes. Bailly and Lavoisier, who were speaking for the government, argued implicitly that scientists alone are capable of passing credible judgment on the mesmerists' claims. Furthermore, and this is crucial, the kind of reasoning scientists employ when doing so is not widely distributed. The truths of nature, according to this image of the scientific enterprise, are not revealed to isolated individuals or validated by popular consensus, they are discovered and interpreted by authoritative figures who have mastered specialized modes of reasoning and are, in almost all cases, members of established corporations. The government's attack on mesmerism was in the hands of agents who spoke for all those who were capable of determining the truth about nature. The general public might find science amusing, instructive, or mystifying, but only those who were, as Condorcet put it, competent scientists could contribute to its advancement.

Bailly and Lavoisier were shrewd enough not to spell all this out in the final report. They wisely demonstrated the authority of science rather than insisting on it, and by carefully limiting the issues addressed, they constructed a report that managed to discredit mesmerism without turning its advocates into martyrs. According to Lavoisier, the committee, after months of weekly sessions around Deslon's tub and after examining all other relevant evidence, unanimously concluded that

what Mr Mesmer and Mr Deslon call the magnetic fluid remained imperceptible and its presence could not be made manifest by any physical experiment. Furthermore, all the techniques by which it is said to be made sensible to sight or touch are illusory and it has no effect on the electric fluid or magnets.[14]

The medical cures the mesmerists claim to effect cannot be counted as evidence. "As Hippocrates tells us, it is nature that cures the sick. The physician's art assists her, but who can distinguish a cure effected by art from one effected by nature?"[15]

To demonstrate that mesmerist crises were caused by suggestion rather than the magnetic fluid, the committee performed a series of experiments with individuals reported to be especially sensitive to the effects of the supposed fluid. When these subjects were not informed that they were near bottles of water that had been highly charged, they failed to react. When blindfolded and told they were being magnetically charged, they reacted powerfully, even when not in fact being magnetized.[16] "These experiments provide a demonstrative proof," Lavoisier wrote, "that imagination in the absence of magnetism produces all the effects attributed to magnetism" and that "magnetism without imagination produces no effects."[17] This mode of testing, which the mesmerists had not anticipated and did not welcome, provided an object lesson for those who failed to distinguish between simple assertions and scientifically demonstrated truths. It also gave the committee precisely the evidence it needed to dismiss as groundless the mesmerists' physical claims.

The committee wished to limit its public critique of mesmerism to the question of physical causation. Breteuil, however, was equally worried about the moral dangers of mesmerism, and he therefore asked the committee to prepare a second report for the king's eyes alone. Was there a danger that mesmerists, men who treated mostly women by touching various parts of their bodies so as to induce emotional crises, might illicitly exploit the vulnerable circumstances in which the patients were placed? The ways in which the mesmerists' modes of treatment and the reactions they elicited might all too easily precipitate a slide from therapy into sex were vividly recounted in lush prose for the king. Arousing scenes and eroticized responses were described in images he must have found as suggestive as baroque paintings of religious ecstasy. The ethical problem addressed was, as indeed it still is, a matter of proper concern to those responsible for the

3 A cartoon of Benjamin Franklin brandishing the report of the royal
commission on mesmerism and the mesmerists fleeing a disrupted
seance.

conduct of medical practice, but in the eighteenth century it remained
a problem that science had not yet successfully appropriated. The
transformation of natural philosophy into the experimental sciences
was then in its early stages and it would be many years before those
seeking to expand the range of scientific discourse would attempt to
make a science out of human sexual relations.

The government's campaign against mesmerism was a complete
success. The beleaguered magnetizers expected that once the committee's report had been published, mesmerism would be officially
outlawed and its practitioners driven into exile, for that was how
contests of this sort had previously been resolved under the old regime.

222

But the government, having learned that it is often less costly to disarm threats than to extirpate them, was satisfied with discrediting mesmerism in Paris. Twelve thousand copies of the committee's report were distributed and the rather creaky machinery of censorship was wheeled out to suppress replies supporting mesmerism. The mesmerists were also subjected to devastating ridicule in *Les Docteurs modernes*, a new play that opened in November 1784, and although they responded to its calumnies as vigorously as they could, they were unable to counter the government's carefully orchestrated public-opinion campaign. By the end of 1784 the Parisian Society of Harmony was splintering into political and spiritualist sects whose mutual antagonism further undermined the credibility of the movement. The Mesmer affair had been great fun while it lasted, but those Parisians who required a steady diet of rumor and scandal now had to look elsewhere for titillating stories of outrageous behavior.[18]

Although reduced to a laughingstock in Paris, mesmerism continued to flourish in the provinces. When it was thought it might be outlawed, an attempt was made to gain legal protection from the *parlements*, but when the threat of repression did not materialize, this improbable alliance between scientific charlatanism and legal conservatism was abandoned. In December 1784 the *Journal de Bruxelles* reported that "if the capital makes merry with the truly comic scenes of the tub, the provinces have taken them seriously: that's where the really heated practitioners are."[19] Naturally the official reports on mesmerism circulated in the provinces as well as in Paris, and the authority of the bodies that issued them carried considerable weight throughout the kingdom. But in the towns and villages far from Paris the claims of those who had been ridiculed by the scientific establishment found a receptive audience among a populace prepared to believe the worst about the privileged few who enjoyed the suspect pleasures of court and capital. A provincial doctor told Vicq d'Azyr that a local mesmerist was building his practice by spreading stories "of the black jealousy felt of [Mesmer] by all the doctors of the capital, of the emotionalism of the Franklins, the Lavoisiers, and all the royal commissioners who had signed the two reports."[20] Mesmerism had been delegitimized in the fashionable world of Paris, but it had not been eliminated throughout

France. When the government faced its own crisis of legitimacy a few years later, the disaffected and embittered defenders of mesmerism returned with a vengeance to settle old scores.

Resentment, although hardly edifying, not infrequently directs events during political and cultural revolutions. Had the French government not collapsed in 1789, we could safely ignore the anger that consumed Jean Paul Marat, Jacques Pierre Brissot, and the other aspiring *philosophes* who railed against the scientific establishment in the early 1780s. But the notorious roles these failed men succeeded in fashioning for themselves in the topsy-turvy world of the Revolution compels us to review more carefully the attitudes they brought to their brief years of prominence. Before the Revolution both Marat and Brissot tried to make careers as scientists and *philosophes*, and like Mesmer, both were driven to distraction when the works they offered as evidence of their merits were rejected as unworthy by those who controlled the high ground of science. Although Lavoisier never showed any particular interest in either Marat or Brissot as individuals or scientists, his prominence as a chemist, academician, and financier was such that he came to represent everything they loathed in the cultural world that spurned them. So long as that world endured, Lavoisier could ignore the disappointment and envy of those who had not succeeded in establishing themselves as full citizens in the republic of science. But his position suddenly became exposed when during the Revolution the social conventions that had protected him were, like the walls of the Bastille, abruptly pulled down. It was then that resentment succeeded in cornering its prey.

Marat and Lavoisier were both born in 1743 and the scientific questions they investigated in the 1770s and 1780s overlapped considerably, yet their encounters with the social world of science were radically dissimilar.[21] Marat came from a small town near the Swiss city of Neuchâtel. His father was a well-established artisan who provided his studious son with a good education. After completing his schooling in 1760 Marat spent two years as a tutor in Bordeaux. Evidently he was already familiar with and admired the writings of Jean Jacques Rousseau. Having decided to pursue medicine, Marat spent the next three years living in Paris while attending courses and learning to be a doctor. He moved to London in 1765 and remained in Great Britain for the next ten years. Although he enjoyed some success as a doctor, Marat, like many other aspiring *philosophes*, soon began styling himself a philosophical physician. In 1775 he obtained a medical degree from the University of Saint Andrews in Scotland without ever visiting the city and the following year he was back in Paris.

While in England Marat acquired a taste for political liberty, but without grasping the interdependence of civil conduct and liberalism. In 1774 he published a lavish book called *The Chains of Slavery* in which "the clandestine and villainous attempts of princes to ruin liberty are pointed out, and the dreadful scenes of despotism disclosed." He was then in his thirties and had begun to hone the rhetoric of violence and denunciation he would employ with such telling effect after 1789. He excoriated Lord North's government and presented himself as the scourge of despots, whom he condemned for keeping their subjects in bondage. One way they did so was by patronizing the arts and sciences. Marat's tone was somber, his message severe.[22] Although he continued to seek academic approbation, Marat had become convinced it was the people, not privileged academics and aristocrats, who determined what is true and just.

Once back in Paris Marat established a modest name for himself as a medical practitioner while devoting more and more time to the investigation of fire, light, and color. He devised a microscope with which he cast images of heated bodies in a way that made visible the shimmering heat they give off.[23] These images, he claimed, allow us to observe directly the fluid matter of fire and hence are of great significance in physics. Toward the end of 1778 he wrote up his experiments and submitted them to the judgment of the Academy of Sciences. The two regular members assigned to evaluate his work were the physicist Le Roy and the chemist Sage. Their report was favorable, in that it declared his experiments novel and of interest, and Marat was understandably encouraged.

In June 1779, shortly after receiving this report, Marat forwarded a second treatise in which he described his recent discoveries on the nature of light. Le Roy was once again assigned to the committee charged with evaluating Marat's claims. The committee reported in May 1780 that it had examined some of Marat's many experiments but was not convinced they supported his conclusions. Furthermore, it considered his views on light to be so much at odds with what was well known that it did not recommend that they be sanctioned or published. Marat, rebuffed but undaunted, published his treatise on light independently, saying that he was submitting it to "an enlightened and impartial public: it is to its tribunal that I appeal with confidence, the supreme tribunal whose decrees scientific bodies are themselves forced to respect."[24] In the same year he published his major treatise on the nature of fire.

Marat, like a comet following its own eccentric path through the heavens, made his closest point of approach to scientific respectability in 1780. After that the members of the Academy viewed him as merely another self-deluding petitioner clamoring for their attention. In the face of this dismissive scorn Marat's resentment ripened into implacable hatred. Le Roy had warned him that, given the subjects he was investigating, Marat had to convince Lavoisier that his experiments were accurate and significant.[25] Perhaps Marat approached Lavoisier; if he did, he had no success. In June 1780 Lavoisier told his academic colleagues that Marat was claiming they had been persuaded his experiments with the microscope rendered the substance of fire visible. Since this was not true, the Academy directed Le Roy to see that Marat's claim was publicly denied.[26]

Brissot, twelve years Marat's junior, was still in his early twenties when he arrived in Paris in 1777, determined to live by his wits and his pen.[27] He began writing and publishing on a vast range of subjects while serving what he assumed would be a brief apprenticeship prior to being welcomed into the ranks of the *philosophes*. Like Marat, Brissot knew his Rousseau and esteemed the England described by Voltaire and Montesquieu. When in 1779 he received several thousand livres from his father's estate, a windfall that must have surprised this thirteenth son of a tavern keeper, he moved to London to establish a "Lycée" open to all the world's philosophers. The project soon failed, consuming all of Brissot's legacy and burdening him for years with debts and lawsuits.

Driven into the literary underground of Paris from which he seemed unlikely to escape, Brissot became exquisitely sensitive to the faults of the world of polite culture that remained closed to him.[28] A passage from a tract he penned in 1782 reveals that he had joined the chorus of those who sought to combine concepts of science, republicanism, utility, honor, despotism, and public opinion in a generalized attack on arbitrary cultural authority. The fact that Lavoisier could have subscribed to such a statement would not have reduced Brissot's anger against the royal authority that reinforced the Academy of Sciences' cultural hegemony.

> The empire of the sciences must know no despots, no aristocrats, no electors. It presents the image of a perfect republic, in which the most useful of merits constitute the sole title to honor. To admit a despot, or aristocrats, or electors officially empowered to set the seal on the production of genius, is to violate the nature of things and the liberty of the human mind. It is a crime against public opinion, which alone has the right to crown genius; it introduces a revolting despotism, making each elector a tyrant and turning all other savants into slaves.[29]

Between 1780 and 1783 Marat and Brissot became fast friends while campaigning together against the haughty academics.[30] During these years Brissot championed Marat's scientific claims as if they were his own; he published, for instance, an account of a dispute he stirred up with Laplace over Marat's optical theory.[31] Marat for his part became so infuriated when told the physicist Charles had made scornful comments about his work that he accosted the academician in his lodgings, where the two men came to blows.[32] By the end of 1783 Marat had completely transformed his personal failure as a scientist into a conspiracy theory of science. He had not abandoned his faith in the cognitive authority of science, however, and he continued to pursue intriguing if unconvincing investigations of such topics as the medical use of electricity and the origin of colors. Knowing, however, that he would never find a place at science's high table, he concluded that the social world of science was rotten to the core. His faith in science became that of a true believer rather than that of an enfranchised citizen.

Mesmer, Marat, and Brissot were all convinced they were being persecuted; it is therefore hardly surprising they found it difficult to work together and often turned on one another. And although Marat and Brissot made Rousseauist claims for the political supremacy of public opinion, their protests, like that of Mesmer, were fundamentally diatribes against the established authorities who had found their works unworthy.[33] During the first half of the 1780s these men forged a rhetoric of denunciation that to everyone's surprise later had enormous influence during the Revolution. It was this unanticipated consequence of the Mesmer affair that gave the public examination of animal magnetism conducted by Bailly, Lavoisier, and their colleagues such unexpected historical significance.

The coalition that destroyed the credibility of mesmerism was too habituated to power to treat every instance of opposition as evidence of a conspiracy, yet they knew that Mesmer was not alone in indicting the public authority of leading scientists. In a secret police report written in the early 1780s Marat is called a "bold charlatan" and it is noted that "M. Vicq d'Azir asks, in the name of the Société Royale de Médicine, that he be run out of Paris."[34] In July 1784 Brissot was locked up in the Bastille on suspicion of having composed satires on public officials. But the political liberals who put their faith in science did not feel unduly threatened, and, having demonstrated that they knew how to control public opinion as well as they controlled the institutions of government, they largely ignored the impotent fury of their critics.

This fearless disregard proved dangerous, for the enemies of the reforming ministers were left free to concoct charges that suddenly appeared credible when seen in a revolutionary light. Brissot, after joining the Kornmann group in the Paris Society of Harmony in 1785, published an anonymous pro-mesmerist manifesto titled *A Word in the Ear of the Academicians of Paris* (*Un Mot à l'oreille des académiciens de Paris*). As one would expect, the Academy of Sciences is bitterly criticized for its treatment of Mesmer and Marat. The pamphlet also contains a virulent personal attack on Lavoisier for his failure to acknowledge his scientific debts to others, for the way he used his power in the academy, and for his activities as a member of the Tax Farm.[35] A more infamous and far more damaging personal attack appeared in Marat's well-known inversion of the charge of quackery, the pamphlet titled *Modern Charlatans, or Letters on Academic Charlatanism* (*Les Charlatans modernes, ou Lettres sur le charlatanisme académique*). Although not published until 1791, this work was written many years earlier.[36] In it Marat fulminates against

Lavoisier, the reputed father of all the discoveries one hears of. Since he has no ideas of his own, he appropriates those of others, but not knowing what to make of them, he abandons them as readily as he takes them up, changing systems as easily as he changes his shoes. During just six months I have watched him take up in turn new doctrines on the principle of fire, on the igneous fluid, and on latent heat. During an even shorter period I have seen him gorge himself on pure phlogiston and then prosecute it mercilessly. Proud of his elevated accomplishments, he rests on his laurels, while his hangers-on praise him to the skies.[37]

Knowing the fate that awaited Lavoisier, one shudders when reading such angry ranting. Yet we should not make the mistake of thinking that this invective is energized by a particularly well-grounded critique of pre-revolutionary science. For Marat, far from constructing a prescient vision of science and democracy, fantasized a people's science with which to pillory those who had rejected him. In this he, like much of the Revolution to which he gave voice, was rebelling against modernism rather than anticipating its farther reaches.[38] Had his self-regard been less offended and had he been more favorably disposed to the possibilities of gradual liberal progress, Marat might have responded sympathetically to the more moderate critique of science developed by the bourgeois mesmerist, Nicolas Bergasse. And had he defended mesmerism as Bergasse did, Marat might have ended up as Lavoisier's ally rather than his sworn enemy.

Bergasse is interesting because he championed a liberal and meritocratic organization of science.[39] One of five sons of a prominent commercial family in Lyons, Bergasse had the good fortune to be spared Marat's and Brissot's anxiety over status and income. He and his brothers, Bergasse wrote, shared the profits of the family enterprise "in a kind of republican arrangement amongst ourselves, all our wealth being held in common and no one desiring to be richer than any other." From this comfortable platform he proclaimed the virtues of ambition and merit while decrying hereditary aristocracy and privilege.

A co-founder of the Society of Universal Harmony, Bergasse objected to the Mesmer committee's 1784 report in terms that expressed perfectly his bourgeois faith in the beneficial results of competition. "What a source of power is ambition," he marveled. "Happy is the state, where, in order to be first, it is necessary only to be greatest in merit . . . Our liberty must be given back to us; all careers must be opened up to us." How much better to live in a society ordered by merit than to struggle in "the sad chaos of feudal government." When academicians exhibit intolerance toward new scientific ideas, it is because they have been granted hereditary privileges. "In general, all exclusive privileges are favorable to some sort of aristocracy; only the king and the people have a constant common interest." In saying this Bergasse was speaking with the authentic voice of the third estate, whose ranks Lavoisier joined even before the revolution had begun. But it was a voice that Marat, by then deeply embittered and committed to denunciation and violence, had learned not to trust.

As official control over the printing and distribution of all forms of literature weakened, the attacks mounted by those who claimed to be spokesmen of the general will grew increasingly strident in tone and telling in effect. These rumbling and complaining voices from below created and then served a new audience that in time learned to make common cause against a disparate array of functionaries they identified as enemies. Those seeking the elimination of hereditary privileges found themselves allied with those opposed to progressive change, while critics of the institutions of science found themselves standing shoulder to shoulder with those who routinely rejected arguments grounded on reason and evidence. Before the collapse of the old regime there was little reason to mount a campaign to convince the governed that they should accept and approve the policies of their governors. Secure in their sense of their own political legitimacy, Bailly, Lavoisier, and their colleagues concentrated on making French public administration conform more closely to the laws of nature. Meanwhile, the spirit of resistance that ultimately brought about a constitutional crisis continued to gather force in the fetid underworld of French political culture.

The story of the ill-fated tax farmer's wall nicely illustrates how a particular complaint came to be blended into this swelling cacophony of opposition.[40] No one enjoys paying taxes and Louis Sébastien Mercier was being melodramatic rather than original when he wrote, in 1782, that he could not walk past the headquarters of the Company of Tax Farmers without being consumed by a desire "to reverse this immense and infernal machine, which seizes each citizen by the throat and pumps out his blood."[41] The year after Mercier scribbled these lines Lavoisier was appointed to the central administrative committee of the Company of Tax Farmers, one of his new responsibilities being to supervise the taxes collected at the various points of entry into Paris. The facilities placed under his command for this purpose were archaic and nearly inoperable, for Paris had long before grown beyond the broken-down remnants of its outermost wall. To make the collection of taxes more efficient, Lavoisier proposed that a new wall be built around the city. Unhappily for Lavoisier, his remedy was widely interpreted as an unjust restriction on the freedoms enjoyed by the citizens of Paris rather than as a progressive contribution to more rational administration.

The taxes imposed on goods entering Paris had grown in a typically helter-skelter fashion over the centuries. In 1719, however, they were reorganized according to eighteenth-century notions of financial rationalization, with the various sales taxes collected by the General Farms being brought together into a single set of Paris entry taxes (*entrées de Paris*). Although vastly simplified, the new system was far from simple. The city was divided into twenty-nine different jurisdictions, each being under the direction of a different office. Goods had to enter the city through specific tollgates, each of which was authorized to collect taxes on different types of commodities. Although over 1,200 agents were employed to collect these taxes, smuggling was widespread. It was estimated that in the 1780s 20 percent of the goods entering Paris evaded taxation and that the annual revenue loss was in the order of 6,000,000 livres. The case for building a new customs barrier, which would be Paris's seventh wall, was compelling. Lavoisier suggested that the cost of construction be financed by the Company of General Farmers. Ultimately, of course, the royal treasury would pay for the wall as a capital improvement for the General Farms.

Construction of the wall began in 1783. This ambitious undertaking, like the other monumental projects that have shaped the architectural fabric of Paris over the ages, was not only to serve a specific function, it was also to be a notable contribution to the continuing reconstruction of the city itself.[42] The architect was Claude Nicolas Ledoux, the design was classic and handsome, the cost eventually exceeded the tax revenues lost in a single year.[43] The wall was built of heavy masonry to a height of about six feet. The sixty-six wooden tollgates were surmounted by highly ornamented stone archways. A continuous patrol road ran along the inside of the wall, while the outside was encircled by a broad boulevard. By 1789 the wall was nearly finished and all but eight of the gates had been installed.

Administrators can initiate projects, but they have little control over how they will be perceived. The meaning of the wall Lavoisier built as a contribution to efficient public administration and as a civic monument was soon being intensely contested. The ornate tollbooths were called fortresses whose real purpose was to exercise control over the city; it was said that the wall turned Paris into a prison; its cost was condemned as extravagant at a time when crop failures and hunger were widespread. A pamphlet published in 1787 condemned the wall for preventing the dissipation of noxious vapors produced in the city, a charge whose multiple ironies must have provided Lavoisier with a few moments of grim amusement. The same publication quoted with evident approval the suggestion, attributed to a prominent French general, that Lavoisier, as the author of this project, should be hanged.[44] The struggle for hegemony over public opinion in pre-revolutionary France was not a game for the faint-hearted.

Public criticism of the tax farmers' wall did not bring down the government, but once the Revolution had begun, the prolonged campaign against the wall helped focus the anger of the Parisians, much as over two centuries later a very different wall in Berlin focused the pent-up political anger of the East Germans. On 12 July 1789, two days before the seizure of the Bastille, furious mobs attacked the new wall, burned several of its gate-houses, and began demolishing the entire structure. Collection of Paris entry taxes ended as resentment found an outlet in action. The public that five years later would witness without regret the summary trial and execution of the tax farmers had spoken; the message it delivered was unambiguous.

Were there any points of intersection between the concurrent debates over the conceptual and theoretical content of science and these strenuous confrontations over legitimacy, authority, and public opinion? Certainly Marat was convinced that his theory of fire and combustion was correct and that Lavoisier's explanations of these phenomena were accepted solely because of his commanding social advantages. And it is true that in the nineteenth century Lavoisier's caloric theory, which asserts that the cause of heat is a substance conserved in chemical reactions, was shown to be false. Does this mean Marat was in fact right when he insisted Lavoisier's theory of caloric was no better grounded than his own theory of fire? Had Marat, through personal pain and social critique rather than by experiment and reason, reached the same conclusion as those twentieth-century theorists who assert that what passes as knowledge in science is in fact socially constructed? We cannot here take up this complex question in its full generality, but what can be said with certainty is that Lavoisier would have been unmoved had the question been put to him. He knew that the superiority of his theories to those proposed by Marat and others could be demonstrated. Truth, for Lavoisier, was ultimately grounded in experimental facts rather than collective opinion.

One wonders what thoughts filled Lavoisier's ever-active mind as he endured enforced idleness while sitting around the mesmerist tub during the summer of 1784. A year earlier Laplace had read their important "Memoir on heat" to the Academy, and in the same month Lavoisier had announced the synthesis of water. The "Memoir on heat" provided a well-developed experimental and quantitative argument for Lavoisier's caloric theory, and following his success on this front, Lavoisier turned his thoughts back to the theory of combustion he had presented to the Academy in 1777. Thus during the summer of 1784 Lavoisier was mulling over the implications of his caloric theory for his comprehensive theory of oxygen. He may have already begun to draft the memoir he called "Reflections on phlogiston," which he read to the Academy during the summer of 1785. It therefore seems probable that as he sat with his knees against the tub and his hands joined with those of his neighbors while awaiting the effects of the magnetic substance, his thoughts fastened on the problem of how best to purge the science he loved of the many insupportable theories that plagued it.

Although Lavoisier's "Reflections on phlogiston" was in no way indebted to the claims of Mesmer or Marat for its conceptual content, it does seem likely that the tone and style of this rhetorical masterpiece reflect a determination to demonstrate to the advocates of ill-founded theories of subtle fluids and substantial fire how good science is done. Brilliant, utterly confident, assertive, and unrelenting in its criticism, Lavoisier's memoir set out to demolish phlogiston by attacking it with a ferocity that was as out of place in the Academy as it was out of character for its author. Having earned the respect of his colleagues through years of careful experimentation, Lavoisier decided to wield his authority with purifying vigor. He would take no hostages. The advocates of phlogiston and other indefensible speculative theories would be harried out of chemistry. The modes of investigation and reasoning employed in experimental physics would then reign supreme in chemistry as well. This *coup de science* would discredit the defenders of phlogiston in the same way that the strategy followed by the committee on mesmerism discredited those who believed in the magnetic fluid. The rhetorical links between Lavoisier's "Reflections on phlogiston" and the final report of the committee on mesmerism thus seem plausible on both psychological and chronological grounds. In both cases the trick was not to banish the opposition but to discredit it. By controlling the opinion of the appropriate public, one could continue to exercise authority and maintain cultural hegemony.

Part III

Revolutionary Politics: 1789 to 1794

10

Representation, Legislation, and Finance

On Tuesday, 4 September 1787, Lavoisier and his wife left Paris to go to Orléans, the administrative capital of the region in which their farm at Fréchines was located. The recently created provincial assembly for Orléans was to hold its first meeting two days later and Lavoisier, having been selected as a representative for the district of Romorantin, would be sitting as a member of the third estate.[1] The king had chosen the first twenty-five members of the assembly: six representatives for the clergy, six for the nobility, and twelve for the third estate. The Duke of Luxembourg, who had extensive land holdings in the region, was appointed president. The members appointed by the king were to elect an additional twenty-four representatives, with their numbers being similarly distributed among the three estates.

The initial session of the provincial assembly was convened with appropriate solemnity. The royal intendant, acting as the king's commissioner, came to the hall where the assembly was to deliberate to officially open their meeting. He was received at the head of the stairway by four representatives, one from each of the first two estates and

The Metropolitan Museum of Art, purchase, Mr and Mrs Charles Wrightsman Gift, in
honor of Everett Fahy, 1977 (1977.10)

4   The 1788 portrait of Monsieur and Madame Lavoisier by Jacques Louis
David.

238

two members of the third estate, Lavoisier and the mayor of Orléans. After being seated the intendant informed the assembly of the king's wishes and the topics they were to address. Following a brief reply by the president, the intendant retired with the same formality with which he had been received.

The assembly met again the following Monday to begin electing its additional members and establishing the rules and schedules to be followed during its deliberations. This preliminary work was completed by the end of the week and the assembly prepared to go into recess until 17 November. Following a formal dismissal by the intendant, the Lavoisiers hurried back to Paris, where many commitments awaited their attention. One of them was to sit for the magnificent portrait by Jacques Louis David that now hangs in New York City's Metropolitan Museum of Art.[2]

What importance should we attach to Lavoisier's participation in the provincial assembly of Orléans, to which, as we shall see, he gave a great deal of time and attention throughout 1788? Was the creation of this novel political forum merely a momentary spasm in a mortally ill body politic, a sign that France was becoming ungovernable and that the possibilities of reform had been exhausted? It is certainly tempting, given the brief history of the provincial assemblies, to hurry past this venture in regional representation. Doing so, however, would seriously foreshorten our understanding of the Revolution itself and, more particularly, of the process by which representative assemblies, which were originally convened to express the views of the people, were transformed into institutions empowered to legislate on behalf of the people. For as Keith Baker has emphasized,

> it was in the context of the great debate over the institution of provincial assemblies, and well before the actual calling of the Estates General, that the French public began to address issues concerning the precise organization of representative assemblies: property versus privilege, as the principle of representation; election versus cooptation, as the means for the selection of representatives; vote by head versus vote by order, as the basis for their deliberation. The social theory of representation therefore lay at the heart of the first full-scale public discussion of the actual principles and procedures according to which the French might participate in the government of a modern society.[3]

The reasons for creating provincial assemblies, and the history of attempts to do so, were matters with which Lavoisier, as a disciple of Turgot and a friend of Dupont, was intimately familiar. The physiocrats were the first to formulate political arguments for creating representative assemblies within a system of governance in which the monarch had absolute authority.[4] In the 1750s Mirabeau wrote of the need to consult the natural orders of society when seeking to involve local elites in the responsibilities of governance. One of the main motives for doing so was the need to check the administrative despotism of royal intendants. During Turgot's ministry in the 1770s, articulating social interests and involving representative groups in the processes of royal administration were matters of official concern. Dupont and Turgot worked together on a proposal they intended to present to the young king, one that called for the creation of a system of hierarchically arranged representative councils. Nothing came of this initiative in the 1770s, for Turgot was dismissed before a serious attempt could be made to implement these ideas. Their plan, however, soon took on a life of its own.

The proposal drafted by Dupont and Turgot addressed the nation's constitutional problems with memorable directness. They told the king that the intractable political contentions besetting France arose from

> the fact that your nation has no constitution. It is a society composed of different orders badly united, and of a people in which there are but very few social ties between the members. In consequence, each individual is occupied only with his own particular, exclusive interest and almost no one bothers to fulfill his duties or know his relations to others. It follows that there exists a perpetual war of claims and counter-claims which reason and mutual understanding have never regulated, in which Your Majesty is obliged to decide everything personally or through your agents.[5]

To move beyond this debilitating war of all against all, the nation needed institutions that accurately represented the natural interests of its people. By deliberating together, the members of these councils can articulate the needs and concerns of the people and convey them to royal officials. In this way the sense of involvement and participation that is fundamental to the creation of a civil society can be developed. The possibilities, Turgot assured the king, were positively Edenic: "You could govern like God by general laws if the various parts composing your realm had a regular organization and a clear understanding of their relations."[6]

The logic of this vision is administrative rather than legislative. Turgot and Dupont, and somewhat later Condorcet, were not suggesting that any of the proposed representative councils and assemblies, from the municipal to the national, be given the power to govern; the king's ministers would continue to be responsible for drafting and implementing the nation's laws. The representative bodies would perform informational and administrative functions. They would make known to the king and his ministers what problems needed to be addressed and how they might be resolved; they would also convey to the people the meaning of royal decrees and their rationale. A hierarchical system of representative assemblies would articulate the interests of every member of society. Those having administrative responsibility could then determine what courses of action would best serve the needs of the nation and its citizens. Public administration would essentially be depoliticized. The king would continue to reign, but the logic of the system and its political legitimacy would be grounded on its being representative and rational rather than on its historical or divine sanction.

This innovative approach to the problem of representation was not much favored during the ten years of conservative reaction that followed the dismissal of Turgot, but the steadily worsening financial situation eventually forced the monarchy to reconsider it. Necker in the early 1780s, and Calonne after 1786, tried desperately to involve local and national notables in public administration without at the same time validating the ancient opposition of the aristocracy, the Church, and the *parlements* to the consolidation of royal authority. New taxes had to be imposed or the state would go bankrupt, but as the recently completed American revolution had made plain, taxation without representation was politically insupportable. Necker, while first minister, therefore took some tentative steps toward creating a few regional consultative assemblies. And Calonne, when drafting the reform proposals he presented to the king in 1786, asked Dupont to draw up a comprehensive program for creating provincial assemblies in all the *pays d'élections*, which included all of central France.[7] It is hardly surprising, therefore, that the guidelines under which the Orléans assembly was created were essentially those proposed by Turgot and Dupont in the mid-1770s.

In 1787 Calonne convened a national Assembly of Notables, his intention being to legitimate policies that would enable him to raise desperately needed funds. The members of the Assembly were not, however, inclined simply to approve all the proposals placed before them. While responding favorably to the creation of provincial assemblies, the Notables insisted that they be organized into the three traditional political estates, with the first estate representing the Church, the second the nobility, and the third all other citizens of France. They did, however, accept the doubling of the representation of the third estate and voting by head, so long as the president came from one of the first two estates. Calonne's long-range plan was to establish these assemblies on a firm foundation before calling a national meeting of the Estates General. If this were done, delegates to the national assembly could be selected from the provincial assemblies, rather than from among opponents of royal governance who had not directly engaged problems of public administration. This then was the perfectly reasonable and politically plausible plan for reform through representation to which Lavoisier committed a great deal of time and energy in 1787 and 1788.

During the intervals between meetings, the work of the provincial assemblies was to be carried out by permanent executive committees, and from the outset Lavoisier dominated the executive committee of the Orléans assembly. The surviving records are peppered with evidence of his active and purposeful involvement. He drafted proposals and reports on the refinancing of debts, on the provision of welfare, on surveying the mineral resources of the province, on agriculture and commerce in Orléans, on the collection of taxes, on navigation on the Loire, on the conversion of mandatory road-work (the *corvée*) to a monetary duty, and on wool production in Orléans.[8] He worked assiduously to create consensus within the assembly, modifying his proposals and revising his drafts as the alliances among his fellow representatives shifted to and fro. Apparently he was not bothered by the restriction that the assembly take up only those topics assigned to it and that it limit itself to proposing policies to the government. The authority Lavoisier sought flowed from factual knowledge and practical reason, not from the assertion of individual or corporate rights. The assembly of Orléans offered him a welcome opportunity to demonstrate how a representative body could contribute to beneficial reform in the administration of public affairs. Serving on the executive committee obliged him to spend a great deal of time in Orléans during months in which he had much to do in Paris as well, but clearly he considered the time well spent.

Critics of the royal government had long called for a meeting of the Estates General, but the king's ministers rightly feared that this national assembly, which had not met since 1614, would be dominated by the traditional opponents of a strong monarchy. They therefore stalled for time, their hope being that when the provincial assemblies were well-established, they could draw supportive national representatives from them. But this strategy required more time than was available to the financially beleaguered government, and on 8 August 1788 the first minister Brienne, in an attempt to gain additional financing, announced that the Estates General would be called and that their first meeting was to be held on 1 May 1789. Although the die was cast, the day was not saved. In less than three weeks Brienne was dismissed and Necker was once again pressed into service as the king's first minister. France was clearly on the brink of a constitutional transformation. The first questions to be answered were how should the Estates General be constituted and should its members vote together by head or separately by estate.[9]

This issue was vigorously debated both within the government and by the larger public during the closing months of 1788. The weakness of the royal government and Necker's unwillingness to confront its opponents reinforced the widely held perception that the long struggle between the aristocratic defenders of caste and privilege and the ministerial champions of centralization and rationality was about to be settled. This was a contest in which Lavoisier had been centrally involved for several decades and he was not inclined to witness its denouement without taking a stand. The vehicle he used to state his position was a memorandum that appears to have been prepared for Necker's consideration.[10]

In the longer view, this debate over the composition of the Estates

General occupied only a passing moment at the beginning of a cascade of events that soon overwhelmed the specific reform proposals of Turgot and his disciples. Yet this was an important moment nonetheless, for in the last months of pre-revolutionary civil discourse, Lavoisier felt free to express his political views without being especially concerned about how they would be perceived by an increasingly truculent and threatening populace. Lavoisier's memoir on the Estates General thus provides a concise account of how, at the end of the old regime, he thought France's constitutional crisis should be resolved. The political doctrines that inform this memoir were also those Lavoisier had in mind later on, when he was struggling to make sense of and respond to the revolutionary developments that were overwhelming reactionaries and reformers alike.

While Lavoisier's memoir on the Estates General incorporates Turgot's and Dupont's call for a hierarchy of representative assemblies, it differs fundamentally from the physiocrats' program in its view of politics. Turgot, Dupont, and Condorcet, who in 1788 also addressed the question of how the Estates General should be constituted,[11] called for a system of representation that would replace the received political order of hereditary castes and privileged corporations. Their system, being rational, would also be natural, and being natural, it would not be dependent on any external sanctions or political agreements. Although the reforms they proposed would be instituted within a monarchal state, they would have the autonomous legitimacy of reason itself. Turgot was especially sensitive to claims that his proposals were not self-validating. He insisted the programs he championed did not depend on prior ideological commitments. Distancing himself from the free-thinking *philosophes*, he protested that "I am not an Encyclopedist, because I believe in God." And he distanced himself from the *physiocrates*, who were also called economists, by rejecting their assumption that their policies could be put into effect only by an enlightened despot: "I am not an Economist, because I don't need a king."[12] The authority of the administrative system he proposed was grounded in its foundation in science; as an administrator Turgot was fundamentally unconcerned with politics and power. The constitution that interested him was the structure of society and its representation, not the legal constitution that limited and directed the exercise of political power.

Turgot's supreme confidence in reason shaped his reactions to the political arguments of his contemporaries. Being himself completely devoted to the public good, he was untroubled by Rousseau's championing of the political claims of the people. He praised Rousseau's *Social Contract* for distinguishing between the sovereign and the government: "that distinction encompasses a brilliant truth, and seems to me to render imperishable the idea of the inalienable sovereignty of the people, regardless of the form of government."[13] Turgot's disdain for politics led him to oppose Montesquieu's commendation of the French tradition of mixed government and separation of powers. Setting one branch of government against another struck him as a regressive and irrational legitimation of privilege, and he criticized both Montesquieu's account of the balance of power in traditional France and the British constitutional tradition as it had evolved in England and America. Turgot conceived of the nation as a collection of individuals, not as a collection of contending corporations; his tasks as an administrator were to learn what the interests of those individuals were and to formulate wise policies that would satisfy them to the greatest extent possible.

Lavoisier was indisputably one of Turgot's disciples, yet he did not share his mentor's dismissive attitude toward politics. He too was prepared to acknowledge the political significance of the will of the people and the authority of reason against the claims of the traditional castes and corporations of France. Happily, Lavoisier argued, history did not dictate how the Estates General should be constituted or how it should vote. How these questions of the utmost consequence were answered, Lavoisier wrote, would determine the future "of one of the foremost empires of the world."[14] It is worth noting that at this crucial moment in French history Lavoisier chose to refer to France as an empire rather than a republic. One wonders what led him to shift his imagery forward to a later stage of Roman history. Did he, with Burkean prescience, foresee that the political turmoil in which France was embroiled could be resolved only by establishing an empire of the sort that Napoleon later forged? He insisted, in any case, that the two most important principles to keep in mind when thinking about the constitution of the Estates General were, "(1) that the Estates General be truly representative, and (2) that they be organized in a way that will ensure, in so far as is possible, the greatest good for all those represented."

Lavoisier was primarily interested in the political constitution of France and the ways in which it had been shaped by circumstance.

> The government of France has never had a fixed and settled constitution; the limits of its various centers of power have never been defined. These limits have instead been established by practice as now one and then another group has prevailed in situations in which circumstances have been more or less favorable and in which different groups have known how to make the most of them.[15]

He then applied this practical mode of analysis to a brief review of the evolution of the French constitution. The *parlements*, for instance, once represented the nation, but they later became nothing more than bodies of "legal officials appointed by the king to administer justice to his subjects." This being so,

> it is essential that they be replaced by one or more bodies that are truly representative and have the power and authority to consent to or reject laws proposed by the king, for it is contrary to the principles of monarchy and, more generally, to all constitutions that are not arbitrary to say that a nation can be forced to abide by a law that does not suit it and to which it has not given sanction or assent.[16]

This is a bold political assertion, one that clearly indicates that Lavoisier's political vision was not rigidly bounded by administrative goals and means. Perhaps Franklin had convinced him that the British parliamentary tradition, with its system of internal checks and balances, had much to recommend it, for this is the role that Lavoisier would have representative assemblies play. But in adapting this concept of representation to French circumstances, Lavoisier gave it a universality that Rousseau and Turgot would have applauded.

> The *parlements* . . . are no longer representative because the people, that is to say that part of the nation that is most numerous, most hard-working, suffers most, and is, from these points of view, the most respectable, are

not consulted. When it is a question of representing the nation, the least individual has rights just like the foremost.[17]

Lavoisier insisted that since the Estates General of 1614 was as unrepresentative as the *parlements*, that precedent should not be followed when electing the Estates General of 1789. In the absence of any useful historical guidance, he concluded, reason and fundamental justice should prevail.

> We will not be guided by what our fathers have done, for they acted badly. We will not simply replicate old abuses. The age of enlightenment (*le temps des lumiéres*) has arrived and today one ought to speak the language of reason and reclaim the imprescriptible rights of humanity.[18]

The radical thrust of Lavoisier's political vision, and in particular its linking of representation and legislation, are striking when compared to both the administrative vision of the physiocrats and the conservative vision of those who supported the aristocratic reaction. In a section titled "on the authority that the Estates General possesses by right" Lavoisier wrote that

> we must dare to say it: the legislative power does not reside in the king alone, but in the agreement of his will and that of the nation. The king and his ministers have acknowledged this principle concerning taxes. They have agreed that no subsidy or subvention can be raised until the nation has consented to it; in effect they acknowledge that to act otherwise would be to attack the sacred right of property.
>
> But what's this? The king himself acknowledges he cannot impose a law that touches the least property of his subjects, yet he can issue laws at his pleasure that dispose of their liberty, their honor, their lives! What good is a law that respects property if one crushes under foot other rights no less sacred and much more important?
>
> It should be acknowledged that whether a law is proposed by the king and consented to by the people or proposed by the people and consented to by the king, legislative power in its entirety rests with the Estates General presided over by the king. This august assembly not only has the right to refuse taxes and to submit fruitless lists of complaints, it can also examine the body of law and effect such reforms as might improve it. It can also, in addition to acting on taxation, make general rules regarding legislation, police, and commerce.[19]

Necker may have been appalled that a member of his inner circle advanced such radical ideas, but Lavoisier had correctly anticipated the demands that would find voice in the Estates General and lead to the third estate's seizure of power at the time of the Tennis Court Oath and much that followed. Yet it is also important that the limits built into Lavoisier's proposal be acknowledged. He was, after all, an experienced administrator, and if he recognized the need for new modes of political legitimation, he also knew that the day-to-day operation of the government could not be left in the hands of a representative assembly. Lavoisier's political conception of governance, unlike the sociological vision of the physiocrats, required the presence of a king or a comparable executive. For while the king could not legislate without the consent of the people, as given by their representatives, his executive authority remained undiminished.

> As the king alone has undivided executive power, he alone, after having obtained the necessary sanctions, can publish and implement the law, supervise its execution, and determine when it has been violated and punish those responsible.[20]

Although the king would maintain executive power, his exercise of that power, in addition to being limited by the need for legislative assent, would be constrained by recognition of the fundamental rights of those elected to represent the people. Before convening the Estates General, Lavoisier wrote, the king should therefore promise to respect three specific rights of the assembly.

1. The king must solemnly promise that there will be no secret orders (*letters de cachet*), no orders of imprisonment, exile, or exclusion from the assembly decreed against any members elected to the Estates General . . . Where force reigns, liberty perishes . . .
2. Everything said in and done by the Assembly of the Estates must be printed and made public, and the press must be completely free to publish memoirs and observations concerning administration, politics, legislation, etc. . . . It is only by having a free press that one can foster continual communication between the nation and its representatives and generate a body of ideas, a general opinion, that reflects particular opinions. This is how one can know, so to speak, the will of the nation . . .
3. It must be decreed and declared in advance that the Estates General

will be convened every three or five years, which is to say at an interval set by the Estates themselves, with the concurrence of the king.[21]

Having stated these remarkably liberal constitutional principles, Lavoisier turned to the details of how the members of the Estates General should be chosen, and in this he followed the plan formulated by Turgot and Dupont in the 1770s. It would be best to have four levels of representative assemblies, from the municipal to the national, with the assembly at each level selecting those members who would represent it at a higher level. Lavoisier realized that such a process of selection and representation would not conform to the pattern employed in the 1614 meeting of the Estates General, but that did not constitute a problem. "Today we are more enlightened on the rights of man and peoples and on political constitutions. Hence, when seeking to do what is right, we should not follow the path that in times past others were allowed to follow when doing evil."[22] The real danger lay not with the king but with the privileged orders. "What enlightened people should fear most profoundly is the aristocratic inclination to form a confederation among the higher orders of the state, an inclination that tends to lead to anarchy and ultimately to a feudal despotism."[23] Given the circumstances, Lavoisier concluded, it would be best to abide by the principles adopted by the provincial assembly of the Dauphiné at the meeting it held at Vizille on 21 July 1788. These principles urged that the number of delegates representing the third estate be double the number representing the first two estates, that all representatives be elected, and that voting be by head.[24]

While the political position Lavoisier articulated just before the beginning of the Revolution is notable in itself, it also suggests certain ways in which his views on the relations between science and administration differed from those of the physiocrats. Turgot and Dupont, like other members of the first generation of physiocrats and economists, acquired their understanding of what science is during the decades of the high Enlightenment. Like Buffon and many other early empiricists, they understood the importance of beginning with facts about nature, but they also trusted intuition and speculative reason to a degree that seemed extravagant to experimental scientists like

249

Lavoisier. While publicly rejecting the deductive spirit of system employed by Descartes, these classic *philosophes* exhibited a startling faith in the systematic possibilities of induction. Turgot, for instance, believed his practice as an administrator would be guided by a science of man grounded on an understanding of the individual rights and interests of citizens. "These rights and interests are not very numerous," he airily asserted.

> Consequently, the science which comprises them, based upon the principles of justice that each of us bears in his heart and on the intimate conviction of our own sensations, has a very great degree of certainty . . . It does not demand the efforts of long study and does not surpass the capabilities of any man of good will.[25]

This approach to the construction of a socially useful science of man stands at the wellspring of modern scientific utopianism. Adam Smith, David Hume, Condorcet, and Auguste Comte, among others, each developed this vision in his own way, their common hope being that the proper use of reason, as informed by evidence, would lead to an understanding of society that would eliminate political contention. Because so many of the leading figures in the Enlightenment were captivated by this research program, it has been taken as characteristic of the Enlightenment as a whole. Those who opposed this image of science and its application to society, as did Rousseau and the later Romantics, have therefore been cast as opponents of both reason and enlightenment. This makes for good polemics but bad history. Surely Lavoisier and his fellow experimentalists must also be considered prominent figures in the Enlightenment, even though their notions of how one should go about constructing scientific theories were highly pragmatic and differed profoundly from the views of those who invoked evidence primarily as a point of departure for further speculation. As is so often the case, historical reality is more diverse than its traditional representation. The *philosophes* of the Enlightenment did not speak with one voice on scientific method, on the role of social science, or on the relations between science and politics. The eighteenth century was not as simple-minded as those who revolted against it would have us believe.

Lavoisier always spoke of the sciences in the plural and he considered scientific knowledge in all its forms to be public knowledge. He thought of scientists as members of a republic, an image that, far from being mere rhetorical adornment, expressed his profound appreciation of the importance of deliberation and consensus in the pursuit of knowledge. Truth emerges from polite but intense competition among those men of talent and education who commit time and attention to the subjects in question. Special skills are needed, such as those employed in experimental investigations and mathematical analysis, but the significance of the information they provide is determined in public debate. And since all scientific knowledge is unavoidably limited in range and subject to revision, Lavoisier was not enraptured by the vision of a general science of man. The process by which scientists arrive at the truth is, he believed, inescapably political. He did not, however, believe that the theories constructed by scientists are nothing more than projections of their interests and preconceptions.

If the physiocrats looked forward to a science of man that would replace politics with administration, Lavoisier looked forward to a political culture in which all decision-making would be as open and rational as it is in science. When urging Necker to grant freedom of the press before convening the Estates General, Lavoisier explicitly championed a liberal-democratic theory of public knowledge.

> All members of society, even those who are not representatives or offi-cials, should be asked or at least be allowed to state their views on the major issues under consideration. This collective competition between different kinds of work and advice leads to truth and such perfection as humans are able to attain. The nation will be reassured by knowing that no one would hazard a false induction when the least of its members can at any moment raise his voice and alert others to an error that was committed or an abuse that was introduced.[26]

Science, as Lavoisier understood and practiced it, was incapable of attaining knowledge of such range and certainty that a science-based administration could replace politics. The checks and balances he called for in the governance of the nation were, rather, very like the checks and balances that pertain in the republic of science. Scientists had learned long ago that deliberation and consensus are central to the progress of knowledge; the French constitution also ought to acknowledge the fallibility of human knowledge. Thus although Lavoisier in one sense, like the physiocrats, sought to model politics on science, their images of science differed profoundly. Lavoisier favored a republican, as opposed to a rationalist, model of science, a model that was realized, although imperfectly, in the Academy of Sciences. Unhappily, the political republic he called for, which would have incorporated this liberal view of knowledge, was not realized either in the old regime, during the Revolution, or in its long aftermath.

Necker brought the debate over the composition of the Estates General to a close late in December 1788.[27] The three estates would meet separately, with the third estate having twice as many representatives as each of the other two. When reaching decisions, the assembly was to vote by estate, not by head. It was a compromise that gave something to everybody while leaving the most contentious problems unresolved. Complex regulations governing the election of representatives were published on 24 January. Electoral districts were to be represented directly, rather than through the provincial assemblies, and *cahiers* of instructions and grievances were to be drawn up.

The settlement of these questions brought the nation's constitutional crisis into sharp focus; France suddenly found itself awash in political activity. "The elections of 1789 were the most democratic spectacle ever seen in the history of Europe, and nothing comparable occurred again until far into the next century."[28] Liberal members of the nobility and the swifter members of the third estate were astonished by the opportunities now available to them. There was a general feeling that the time had come to both begin and complete the political regeneration of the nation. The Estates General was not being convened to solve particular administrative problems; the immediate task was to give France a new political constitution. No one quite knew how this was to be done, and in the absence of strong central control, events soon generated a momentum of their own. By midsummer of 1789 it was possible to anticipate what some of the more extreme consequences of this singular exercise in political reform were likely to be.

Lavoisier responded to the calling of the Estates General and the events that followed as one would expect, that is to say by committing himself to more activities than any normal individual could possibly manage. In everything other than physical science he was a man of action rather than theory, and his energy and abilities were such that he was soon busily engaged on many fronts. Although political developments were moving in a direction that made him distinctly uneasy, he seems never to have considered leaving France or withdrawing from public life. He remained utterly convinced that a full and open examination of the many positions he had held would demonstrate he had conducted himself honestly and honorably at all times. He also continued to work for the king and the Academy as long as circumstances permitted, and whenever he could find the time, he returned to his scientific research. Given his pragmatic approach to politics, he never concluded that the situation in which he found himself, no matter how dire, was hopeless. During the early years of the Revolution he did all he could to steer the ship of state away from the rocks; during the later years he did his best to salvage what could be saved from the wreckage.

As an aristocrat and landowner Lavoisier was entitled to vote with the nobility in the electoral district (*bailliage*) of Blois. At the end of February 1789 he journeyed south to join his fellow electors, who numbered slightly less than a hundred, in selecting two delegates for the Estates General and in drawing up instructions on the various questions then under discussion.[29] Inevitably Lavoisier was appointed secretary; by the time the meeting concluded at the end of March he had also been elected to serve as an alternate delegate to the Estates General. When he learned, during his month in Blois, that the city was threatened by famine, he donated 50,000 livres to buy flour for the city's bakers.[30] Although Lavoisier may well have had certain political purposes in mind, this act of charity reveals yet again that even while actively engaged in political reform he remained alert to the pressing problems of his immediate community. Clearly his sense of his duties as a citizen was as informed by a conservative awareness of his individual social responsibilities as by abstract concepts of the rights of man.

The instructions approved by the nobles of Blois embraced many of the principles Lavoisier had already adopted. Indeed, it is remarkable how eager the nobles were, at least in their collective declaration, to eliminate hereditary and class privileges, to acknowledge the rights of all Frenchmen, to introduce an equitable system of taxation, to reform the civil and criminal law, to locate legislative authority in the Estates General, and to emphasize the transcendent autonomy and authority of the nation. During the early months of 1789 the fresh breath of liberal reform sweetened political discourse throughout the kingdom.

Champions of reform often get carried away by their own enthusiasm and the nobles of Blois soon surrendered to this temptation. They called for radical new departures in many areas of society, their optimism arising from their own sense of enlightenment and beneficence rather than from well-developed plans as to how the reforms they embraced should be effected. They were also unrestrained in their determination to crush those who opposed them. Privileges based on class and heredity were declared illegitimate, the rights of man were universalized, the supremacy of the public good was sanctified as absolute. These were the principles that should guide "all the citizens of this great empire."[31] A uniform style of clothing and a uniform system of weights and measures were to be developed for the nation; a national plan of education for all classes was called for; internal trade was to be freed and all internal customs barriers were to be eliminated; the oath taken by soldiers was to be secularized so that they would swear allegiance solely to the king and nation.[32] These reforming nobles had English liberal principles in mind when formulating their political program,[33] yet they obviously were not prepared to show restraint and tolerance in their political practice. When opposing the creation of a French equivalent to the House of Lords, the nobles declared that anyone who dared support such a house, as well as all those who took seats within it, were to be considered traitors to the nation (*Patrie*).[34] Although their goals were admirably liberal in principle, the nobles exhibited an ominous eagerness to compel assent.

Shortly after returning to Paris in April Lavoisier was involved in selecting another set of delegates to the Estates General. The nobles living in the ninth district of Paris, in which the Arsenal was located, were to choose eight electors who would meet with representatives from other districts to select the city's delegates to the Estates General. Lavoisier served as secretary to this group too and, not surprisingly, the instructions they sent forward closely resembled those approved by the nobility of Blois. Lavoisier was also chosen to be one of the eight electors.[35]

The Estates General was convened at Versailles on 5 May 1789 with great ceremony and high expectations. Lavoisier remained in Paris working on his affairs at the Academy, at the Arsenal, and at the Tax Farm during the early months of summer. He kept in touch with developments at Versailles, however, for the momentous events taking place there were being closely followed and intensely debated in Paris. Two intertwined dramas were being played out. The king's ministers, led by Necker, were primarily concerned with immediate problems, the most pressing being financial demands that threatened to force the government into bankruptcy. They also could not ignore the riots arising from widespread food shortages and the need to maintain order within the kingdom. Although the delegates to the Estates General were aware of the government's financial difficulties, they were primarily interested in transforming themselves into a national assembly empowered to refashion the nation's constitution. History has immortalized certain scenes enacted on the grand stage of the royal residence at Versailles, yet at the time the more than a thousand delegates to the Estates General, as well as the chorus of officials, courtiers, and spectators that followed their every move, must have often found the entire spectacle rather bewildering.

Lavoisier was especially interested in Necker's proposals for alleviating the government's credit crunch. In his opening three-hour oration Necker revealed the extent to which the government had already borrowed against anticipated revenues for 1789 and 1790.[36] In the short term he could do little more than continue borrowing, his primary lender being the Discount Bank (Caisse d'Escompte) that Turgot had established in 1776 to provide some of the services of a national bank. Unlike the Bank of England, however, the Discount Bank was a private rather than a public venture. Funded by the sale of shares to a limited number of investors, its main functions were to issue notes that circulated as currency and to discount bills of exchange and other credit instruments for bankers. It was, in other words, another vehicle by which the royal government sought to obtain credit from wealthy French bankers and financiers. By 1787 the Discount Bank was making loans directly to an increasingly importunate government. Thus it came as no surprise during the first month of the Estates General when Necker asked the Discount Bank to grant the government a further loan of 12,000,000 livres. Lavoisier was at the time a member of the Bank's board of directors and at first they resisted, but after receiving further assurances from Necker they agreed to grant the loan. At the end of May the king wrote to the directors, thanking them for their confidence.[37]

The delegates to the Estates General were far more concerned with how they would meet and vote than with the nation's finances. This issue was joined over the verification of election returns, with the third estate insisting that all three estates meet together from the outset. When the nobility and clergy resisted, the third took to calling itself the Commons, and by early June an uneasy stalemate was blocking further action. When several clerical delegates defected to the third estate in the middle of June, the members of the third estate declared themselves the National Assembly. Since the meaning of this claim was unmistakable, the king prepared to call the Estates General to a royal session to reassert his authority. While this session was being planned, the clergy as a whole voted to joined the third estate, but when the two estates appeared at the third estate's designated meeting place on 20 June, they found the doors locked and heavily guarded. They promptly commandeered a nearby indoor tennis court and vowed never to disperse until "the constitution of the Realm and public regeneration are established and assured."[38] When the royal session was finally held a few days later, the third estate, along with its compatriots from the other two estates, responded to the king's assertion that his will was supreme by refusing to disperse. On 27 June, faced with a de facto defeat, the king ordered the first two estates to join the National Assembly. The victory of the third, and the king's acceptance of it, were greeted with delirious public displays of relief and rejoicing. Arthur Young confidently reported that "the whole business now seems over and the revolution complete."[39]

The Revolution had in fact just begun.[40] Despite the king's public gestures, the court had no intention of surrendering sovereignty to an elected assembly. Troops were therefore mobilized and moved into strategic positions, and arms and powder were shifted to more secure arsenals. A plot was also put in hand to dismiss Necker. While these plans were being drawn up at Versailles, the electors of Paris, who had stayed in touch with one another after choosing the city's delegates to the Estates General, began responding to the anxiety produced by troop movements and the royal government's indecisiveness by organizing a city government of their own. The Commune, as they constituted themselves on 13 July, immediately set about forming a militia to keep peace in the city; a few days later they elected the astronomer J. S. Bailly, who had served as first chairman of the National Assembly, as the first mayor of Paris. The Commune also elected the Marquis de Lafayette, the hero of the American revolution, commander of the civic guard, which was promptly renamed the National Guard.

When news of Necker's dismissal reached Paris on Saturday afternoon, 12 July, the reaction was explosive. Lavoisier feared there would be a run on the Discount Bank the next morning, but the truly dramatic action was in the streets. The people of Paris quickly took matters into their own hands. Earlier in the month the citizens of Lyons had protested against high grain prices by destroying the tollgates surrounding the city. The Parisians followed their example on 12 July and laid waste the nearly completed tax farmers' wall. By the evening of the next day hastily thrown together citizen's patrols were in the streets. On the morning of the fourteenth cannon taken from the hospital for military veterans (the Invalides) were placed in front of the city hall and negotiations were underway for the surrender of the Bastille. Almost a hundred members of an impatient crowd that forced its way into the inner courtyard of the fortress were killed when nervous troops opened fire. When cannon brought from the city hall were trained on the Bastille, the commander surrendered and was promptly murdered. The next day the king effectively announced the end of royal authority when he told the National Assembly he was ordering the troops around Paris to disperse. Once again his acceptance of the will of the people was greeted with displays of public affection and relief. Once again it appeared that the Revolution had been brought to a close and chaos avoided.

Lavoisier's hope that France's new constitution would incorporate a system of checks and balances was dashed by the king's fecklessness and the assertiveness of both the National Assembly and the people of Paris. Yet he knew a great deal about ministerial responsibility and administrative complexity, and he realized that effective governance, whatever the form of government, requires more than oratory and legislative mandates. The National Assembly would eventually have to wrestle with financial problems, administrative challenges, and military necessities. At an early stage, therefore, Lavoisier turned away from the fevered political discourse that occupied center stage in this, the first modern revolution, and concentrated instead on the financial and administrative problems that he knew the new regime could not long avoid. He remained, as always, a loyal, talented, and immensely hard-working citizen, his main concern, other than his science, still being, as before, to contribute to the public good.

Lavoisier advised the various governments thrown up by the Revolution on financial matters, but his advice was seldom heeded. In the heat of revolution decisions on the nation's finances were thoroughly subordinated to politics. Lavoisier, however, made no attempt to use his financial expertise to advance any particular political program. Following the momentous events of July 1789, he addressed the Revolution as he found it day by day and described the financial realities as he perceived them to those who happened to be exercising political power. Having devoted many years to studying the nation's economy, he had little difficulty analyzing its problems and possibilities, yet he found it enervating to speak truth to power time and again without having any perceptible effect. In the end he simply had to walk away from a financial situation which had become insupportable. Neither defeated by his detractors nor disgusted by the Revolution, he could no longer effectively serve governments that appeared to be unconcerned with the financial consequences of their decisions. Once the wave of political change had swept past him, he was reduced to assessing events administratively, and this proved to be frustrating. He had lived too long and too well in the orderly world of the old regime to master, much less savor, the storm and strife of a Revolution that refused to end. He was too well versed in economics, which later reformers and revolutionaries would curse as the dismal science, to believe that human society could be transformed by political will alone.

During the third week of July 1789 another attempt was made to bring the Revolution under control. The fiscal situation remained precarious, with the state depending entirely on short-term credit to refinance its enormous debts as they came due. To ensure the continued cooperation of the Discount Bank, the National Assembly, which had recently added the term Constituent to its name, invited the bank's administrators, including Lavoisier, to appear before it on 20 July. They were commended as patriots for having remained open and for supporting their notes during the recent period of unrest in Paris. But the respite was brief. In the countryside the spreading panic known as the Great Fear was gathering force and by the beginning of August the National Assembly was once again vainly struggling to gain control of the genie of revolution.[41]

As events lurched forward unpredictably, the deputies in the Constituent Assembly sought to deflect the mounting fury by jettisoning what remained of the old regime. At the astonishing meeting of 4 August, "the most sweeping and radical legislative session of the whole French Revolution,"[42] the National Assembly in an orgy of renunciation irrevocably abandoned the rights and privileges of caste and corporation that had been central to the social and administrative order of traditional France. Their original purpose had been to calm the countryside, where anxious and angry peasants were achieving by arms what the Assembly sought to legitimate with legislation, but in their intoxication with liberty they went further than anyone had anticipated. By the time dawn broke on 5 August, the nobility had lost their hunting rights, the clergy had lost its right to tithe, towns and provinces had renounced their ancient liberties, venality of office had been eliminated, and it had been decided that every Frenchman would henceforth enjoy complete fiscal and civil equality. Toward the end of the month the *Declaration of the Rights of Man and Citizen* was issued in yet another attempt to bring this grand but dreadful business to a close.

The abolition of feudalism affected the nation's financial situation in two different ways. While there was general agreement that privileges had to yield to rights, there was also a firm consensus concerning the sanctity of property. Feudal dues were therefore to remain in force until those holding them were compensated for their losses, and individuals who had purchased royal offices in return for certain privileges and incomes were to be reimbursed by the state. What the Constituent Assembly failed to notice was that tax collections and the payment of other traditional obligations had already fallen into complete disarray.

Necker, however, knew something about balancing accounts and in September he persuaded the Assembly to call for a "patriotic contribution" of one quarter of every citizen's income. Although Lavoisier agreed to contribute 30,000 livres, the overall results were predictably insufficient.[43] The legislative mandate to abolish feudalism added to an already overwhelming national debt, a debt that was, as Mirabeau put it, a national treasure that had to be honored.[44] But the overthrow of the old regime also made available certain resources, most notably royal estates and church properties, that could be used to pay off the debt and reorder the nation's finances. Yet it was only after the king had been forced by popular agitation in October to accept what had been decreed in August that the Assembly at last came to grips with the financial consequences of the revolution it had initiated.

In the financial debate that occupied the Constituent Assembly in November and December 1789 the flame of politics played over the tinder of entrenched animosities, the result being a fiscal conflagration of truly revolutionary proportions. The central problem remained that of managing the nation's short-term indebtedness. On 14 November Necker proposed that the Discount Bank be turned into a national bank whose notes would be guaranteed by the government but whose funds would be provided by private financiers. Lavoisier, who was then serving as president of the Discount Bank, appeared before the Assembly on 17 November and mingled stern warnings on the need for financial discipline with appropriately fervent assertions of patriotism. The Discount Bank, he reported, would cooperate in the formation of a national bank if requested to do so. Necker's proposal was opposed by two groups, however. One consisted of the established Catholic financiers who resisted the intrusions of foreign Protestants like Necker, who was Swiss, into the previously closed world of royal finance. The other was made up of provincial deputies who feared that the bankers of Paris were maneuvering to capture control of the nation's credit so as to impose their self-serving schemes on the country as a whole. Mirabeau, with characteristically brilliant oratory, united these two camps and led them in a successful campaign against Necker's proposal.

Since the Discount Bank was a creation of the old regime, Mirabeau argued, it would inevitably be an instrument of despotism. He insisted that the nation's finances had been too long in the hands of capitalists interested only in profits, whereas they should be administered directly by agents of the state. Although Lavoisier appeared a second time before the Assembly to defend the Discount Bank's record, the hatred of financiers and tax farmers was such that the evidence he offered carried little weight. To save France from the "bloodsuckers on the body politic" who have reaped fortunes from "the sweat and blood of the people," the Assembly rejected Necker's proposal and refused to create a bank whose credit depended on investments made by shareholders. The National Assembly, which had so recently staked its claim to legislative hegemony, decided instead to use its legislative mandate to control directly the nation's financial destiny as well. The costly and cumbersome services of private financiers were no longer necessary. Mirabeau and the general will had triumphed; the people of France were to be released from their bondage to capitalists.

This uncoupling of public and private credit had predictable, and predictably dire, consequences. Whereas the appetites of the Bourbon kings had been somewhat constrained by their need to secure private financing, the modern nation-state insists that its political legitimacy gives it the authority to compel financial support. In some ways this signaled a return to an earlier era in which kings repudiated debts and seized property when in financial distress, even though such tactics had seldom been used in France in the century before the Revolution. Indeed, if acknowledging that the laws of economics cannot be abrogated by legislative decree is a sign of modern rationality, then the old regime was, at least in this regard, more modern than the governments that succeeded it. But the National Assembly was entirely modern in its insistence that private property be respected and in its refusal to repudiate the national debt. In making the National Assembly directly responsible for the national debt, however, the opponents of Necker's plan were accepting a Herculean task. Their strategy was to pay off the debt by selling the lands seized from the Church. Despite the vast program of nationalization carried out by the Revolution, however, most of the debt was in fact repudiated by that most insidious weapon of modern finance, the devaluation of paper money.

The Revolution's commitment to the protection of private property was understandably not absolute. The supremacy of the general will was interpreted as meaning that individuals and corporations could only continue to hold property so long as they used it to serve the general welfare. This principle of patriotic utilitarianism encouraged revolutionary spokesmen to call for the nationalization of royal estates, of Church property, and of the properties of political émigrés. At the beginning of November the Assembly passed the necessary laws and decreed that administrative control over these lands would remain with the government rather than being turned over to private contractors.

During the last two months of 1789 the Assembly addressed the linked problems of managing both the nationalized lands and the national debt. On 18 November, four days after Necker proposed turning the Discount Bank into a national bank, the finance committee of the Assembly recommended that 400,000,000 livres worth of Church and Crown land be sold. This seemed a modest proposal, for it was thought that the remaining nationalized lands were worth roughly two billion livres. Debate then focused on how the sale should be carried out. Mirabeau campaigned effectively against having private capitalists serve as brokers. He proposed instead that the state issue special interest-bearing bonds called *assignats* that could be used to purchase nationalized lands. This would give the government a new source of credit with which it could pay its outstanding obligations while insuring that the bonds were fully backed by real property and would be retired without having to be refinanced. On 19 and 21 December the Assembly passed decrees rejecting Necker's proposal for a national bank and establishing a special administrative office (the Caisse de l'Extraordinaire) to handle the issue of nationalized lands. The first sale of *assignats* was also approved.

The political situation was still unsettled during the early months of 1790, as the Caisse de l'Extraordinaire was getting organized, and the financial situation continued to deteriorate. In February Lavoisier indicated, in a letter to his old friend Benjamin Franklin, how difficult the situation was becoming for political moderates.

> After having held forth to you about what is happening in chemistry, I should say something about our political revolution. We look upon it as achieved irreversibly. Nevertheless there still exists an aristocratic party which makes futile efforts and which is evidently very weak; the democratic party has on its side the greatest number and in addition education, philosophy, and enlightenment. The moderates, who have kept their

heads in this general turmoil, think that circumstances have carried us too far; that it is unfortunate to have been obliged to arm the common people and all the citizens; that it is ill advised to place force in the hands of those who should obey; and that it is to be feared that the establishment of the new constitution will be opposed by the very persons for whose benefit it has been made . . .

We deeply regret at this time that you are so far from France; you would have been our guide, and you would have marked out for us the bounds that we should not have exceeded.[45]

Necker's position was fatally undermined in March when it was revealed that he had grossly underestimated the shortfall in tax collections. In April the Assembly, grasping at one of the few means available to it, made *assignats* legal tender, issued them in much smaller denominations, and reduced the interest paid on them so as to encourage conversion into land. In the next six months the deficit mounted and six additional issues of *assignats* were authorized. In July Lavoisier told the Scottish chemist, Joseph Black, "the Revolution that is taking place in France must naturally make some of those attached to the former administration superfluous and it is possible that I may enjoy more freedom."[46] By the beginning of September Necker, who understood that credit cannot simply be manufactured, had been hounded from office; by the end of the month *assignats* to a value of 1,200,000,000 livres and bearing no interest had been authorized. The Revolution was now being financed primarily by the printing press.

This slide down the slippery slope of monetary inflation had not gone unnoticed or unopposed. In August 1790 the finance committee had sought guidance on how best to use the remaining national lands to reduce the outstanding debt. During that month Lavoisier addressed the question in a memoir on *assignats* read to the Society of 1789, a prominent association of moderate revolutionaries.[47] Lavoisier did not oppose using nationalized lands to retire the debt, nor did he argue that private financiers should play a role in their sale, but he was concerned about the way the issuing of *assignats* was being carried out. He insisted above all on accurate accounting and fiscal prudence. Essentially there were two problems. One was the disastrous effect on prices and commerce of rapid changes in the supply of money. It is interesting that Lavoisier illustrated this point by quoting at length from the economic writings of the British philosopher, David Hume.[48] The other problem was the slow rate at which *assignats* were being retired by exchanging them for land.[49] To remedy these problems Lavoisier suggested that the terms under which *assignats* were issued be altered in certain specific ways. Despite their economic merits, his proposals did not carry the day and the printing of *assignats* continued unabated. As Simon Schama has noted, "the revolution changed much less in France than we often suppose, and one of the matters in which it did no better than the monarchy was the way in which short-term exigencies controlled longer-term fiscal rationality."[50]

By the end of 1790 Lavoisier could do nothing more than describe developments he had little hope of influencing. In March 1791 he provided the Assembly with a comprehensive account of his estimates of the nation's wealth. A few days later the tax farmers' lease was cancelled retroactively, as of 1 July 1789, and they were ordered to prepare their accounts for inspection. In April he was appointed as a commissioner of the recently created National Treasury, but to avoid the appearance of drawing two salaries, and more particularly to protect his position on the Gunpowder Commission, he declined to accept any payment for his services in the new position.[51]

Hope that the Revolution was at an end and that French political life might finally settle down revived during the autumn of 1791. In September the Constituent Assembly completed and approved the new constitution, and in the same month the king granted his assent. The members of the Constituent Assembly, who had been sitting since the opening of the Estates General, declared themselves ineligible for election to the first Legislative Assembly to be formed under the new constitution. At the end of September the National Assembly was dissolved; the next day the Legislative Assembly held its first meeting. Optimists, and most notably the Feuillant party that had championed the new constitution, rejoiced as France embraced constitutional monarchy. But others realized that just below the surface France was still wracked by potentially fatal antagonisms. As the British ambassador observed, "the present constitution has no friends and cannot last."[52]

In November Lavoisier reported to the Legislative Assembly on the state of the nation's finances. The following month his report was published by Dupont, who had recently set up his own commercial press.[53] Although Lavoisier had long believed that France should be governed by a constitutional monarch, his mood at the end of 1791, as conveyed in the opening pages of his report, was pessimistic.

> At a time when everything is exaggerated, the good as well as the evil, and in which everyone looks at things through instruments that enlarge or diminish them, distance them or make them appear to be close, or in which no one sees things either in their true dimensions or in their proper places, I thought it would be useful if someone attempted to discuss dispassionately the way things are, to submit the finances of the state to a rigorous arithmetical analysis.
>
> It seemed to me that what I have previously done solely for myself might be useful to others, perhaps even to legislators who are going to be engaged in restoring our finances . . .
>
> The reader should expect to find nothing here except what is required by the subject: this document will be as cold as reason. I wish it contained nothing but statements and numbers; I wish I could eliminate all vestiges of reasoning, for facts are the givens that never deceive us – it is in judging that we go astray.[54]

Lavoisier wished the new Legislative Assembly well as it wrestled with problems that would profoundly affect "the state of the Empire, which has been continually battered by a succession of events that evidently cannot be arrested,"[55] but his heart was not in it. Early in 1792 he wrote to a friend that "what is occurring time and again is what can be observed in all popular governments: they know no bounds. One would expect greater peace under a philosophic constitution that was adopted so everyone could live in peace."[56] As the commencement of war approached and tax revenues continued to decline, Lavoisier concluded his usefulness to the government was at an end; in February he resigned from the Treasury.

Lavoisier's integrity and his extraordinary ability to manage complex administrative problems were sorely missed, and shortly after leaving the Treasury he was asked if he would serve as commissioner of public revenues. As he explained to the king, however, in an affecting letter dated 15 June 1792, he could no longer respond to such a call to duty. Lavoisier, knowing full well that the sentiments expressed in this letter would probably be widely reported, wrote with astonishing frankness, considering the political temper of the time.

Sire, I wish to assure you that it is neither craven fear, which is foreign to my character, nor indifference to public affairs, nor a feeling that I lack the necessary force, that obliges me to decline the offer, which honors me by indicating your majesty's confidence in me, to serve as head of the ministry of public revenues. While serving in the National Treasury I had ample evidence of your majesty's patriotic feelings, of your sincere concern for the happiness of the people, of your unwavering rectitude, and of your steady probity. I am therefore aware in ways that I cannot express of what I am denying myself by forgoing this opportunity to become an instrument of such feelings towards the nation.

It is, however, the duty of a upright man and citizen to not accept an important position unless he can expect to fulfill his obligations to their full extent.

I am neither a Jacobin nor a Feuillant. I am a member of no society or club. Being accustomed to weighing everything on the scales of my own conscience and reason, I would never consent to allowing any party to determine my views. I have with heartfelt sincerity sworn to uphold the constitution, which you have accepted, and the powers granted by the people to you, Sire, you who are the constitutional king of the French, you whose hardships and virtues are not adequately appreciated. Convinced as I am that the Legislative Assembly has exceeded its constitutionally defined limits, what should a constitutional officer do? Someone in that position, being unable to act according to his principles and his conscience, would appeal in vain to the law to which all Frenchmen are bound by a most solemn oath. Such resistance as he might suggest, by employing the means which the constitution allows your majesty, would be seen as a crime. He would perish, a victim of his sense of duty, and his inflexible character would become an additional source of unhappiness.

Permit me, Sire, to continue to dedicate my remaining labors and my existence to service to the state in some less important position, one in which my efforts can perhaps be of greater and more lasting usefulness. Being devoted to public education, I will look for ways to instruct the people on their duties. As a soldier and citizen I will bear arms for the defense of the country, for the defense of the law, and to defend the imprescriptible rights of those who represent the people of France.[57]

The constitutional arrangements adopted in 1791 rapidly unraveled as events forced the Legislative Assembly and the king into irreconcilable confrontations. The king's credibility had been severely damaged by his secretive attempt to emigrate in June 1791, the famous flight to Varennes. Shortly after the Legislative Assembly convened, Brissot began calling for punitive sanctions against all émigrés, and when the king vetoed the proposed legislation, as was his constitutional right, his actions were interpreted as an attempt to frustrate the will of the nation. Suspicions about his devotion to the Revolution were strengthened by his veto of a bill imposing harsh penalties on members of the clergy who refused to take the civic oath required of them. These differences might have been compromised had invasion not been imminent, but once war had been declared against Austria in April 1792, reconciliation was no longer possible. By July the people of Paris had once again seized political leadership and firebrands were insisting that the fatherland was in danger. France was on the brink of what would become the first mass mobilization of a modern nation-state.

In August 1792, three years after the abolition of feudalism, Paris was again in an uproar. Lashed on by Marat's bloodthirsty rhetoric, the leaders of the Commune and the Jacobin Club were calling for a new constitution. The Legislative Assembly, having lost all control of events, agreed to the formation of a Convention that would "assure the sovereignty of the people and the reign of liberty and equality."[58] Elections were held early in September, and when it was proven that the king had been engaged in a secret and treacherous correspondence, the demand that the new government be a republic proved irresistible. The French monarchy was formally abolished on 21 September 1792. The next day, later designated as the first day of the revolutionary calendar, France was declared a republic. The matter of disposing of the king was finally taken care of the following January.

Lavoisier was wise to leave the administration of the nation's finances to others. The Caisse de l'Extraordinaire was suppressed at the beginning of 1793, an act which merely acknowledged that the issuing of *assignats* was no longer linked to the sale of nationalized lands. In April *assignats* were made the principal legal currency and as such they were to be used to settle all government accounts, including soldiers' pay. The defense of the country now required that goods and services of real value be exchanged for paper of rapidly decreasing value. The presses worked ever faster to meet the growing demands of a nation in arms. Eventually, of course, the *assignat* became worthless, but only after notes having a face value of over 40 billion livres had been issued. The end came early in 1796 and was acknowledged with exquisite appropriateness by a ceremonial breaking of the presses.[59]

The end of monarchy and the execution of the king finally dashed all hope that the Revolution might culminate in a constitutional settlement embodying a system of checks and balances. Lavoisier had in fact expected the new form of government to be limited in two ways. Politically, he expected its power to be divided between the legislative and the executive branches, each of which would prevent the other from acting despotically. Philosophically, he expected both branches of the new government to pursue policies that conformed to the laws of nature, including those of economics and society, as determined by scientists. Both politics and science had a role to play in the conduct of public affairs; what was wanted was an informed and respectful dialogue between them. But in the maelstrom of revolution this goal proved to be unattainable. The events transforming France did not invalidate Lavoisier's analysis of what was needed, nor did they persuade him to give up on his nation and flee. But they did force him to the margins of public life, where he watched with growing foreboding and dismay as the revolutionary storm grew in fury and destructiveness.

11

The Republic of Virtue

In August 1789, at the very beginning of the Revolution, Lavoisier narrowly escaped being lynched by an angry mob of Parisians. The story of this encounter illustrates several themes that persisted throughout the entire period of revolution: how confusion and fear drove events forward in unpredictable directions, how those charged with public responsibilities frequently responded with courage when threatened with violence, how normal procedures often collapsed under pressure, how social disorder fed on deep currents of suspicion and antagonism, and above all how intimidation and physical force came to dominate French public life. The Revolution was tinged with terror from July 1789 onward. If it created opportunities for national regeneration, encouraged extravagant hopes, and called forth transcendent exertions, it also unleashed a dark undercurrent of chaos that mercilessly terminated the lives of many decent people before finally being contained by the establishment of a new order.

The tension in Paris was palpable at the beginning of August 1789. The Great Fear was raging in the countryside and the city, under its new mayor Bailly and Lafayette's recently founded National Guard, nervously awaited some indication of what would happen next. The

National Assembly had exhausted itself on the fourth in an orgy of renunciation, but it was not yet known how the king would respond to this legislative annihilation of the old regime. To ensure that Paris would not be left defenseless, an order had been issued requiring Lafayette's authorization before any gunpowder could be moved out of the city. Lavoisier at that time wished to send some inferior powder of the type supplied to slave-traders (*traite*) to the powder mill at Essonnes; he planned to replace it with higher-quality musket powder. Not having been able to locate Lafayette, he had the required order signed by his deputy, the Marquis de La Salle. The powder was then loaded on a barge under the watchful eye of four soldiers from the district.[1]

On 5 August people living near the Arsenal became suspicious. Seeing barrels of gunpowder labeled *traite* being taken from the Arsenal, they were easily convinced the powder was being sent to traitors (*traîtres*). A crowd quickly formed and a delegation was dispatched to the city hall. Bailly and Lafayette, not knowing that official permission had been granted, decided the barge should be unloaded until the matter was cleared up. This apparent confirmation of the crowd's fears appeared to validate further rumors that other weapons were being taken from the Arsenal. The crowd then insisted that the hapless soldiers who had been supervising the transfer be thrown in jail and the barge be placed under a close watch.

The following day Lavoisier and the commandant of the district of Saint Louis, in which the Arsenal was located, attempted to placate the angry citizens. The guards who had been arrested were released and Lavoisier showed his signed authorization to representatives of the Commune. Lavoisier and two representatives then went to the Arsenal to calm the crowd surrounding the barge. He opened several barrels chosen at random and explained what kind of powder they contained. The crowd remained suspicious, however, and insisted he test the powder he said was inferior by setting fire to a sample. When it burned they were convinced it was good powder and that Lavoisier was trying to deceive them. They forthwith insisted that Lavoisier and another director of the Gunpowder Administration, the elderly Le Faucheux, return with them to the city hall. They did not disturb a third director, Clouet, who was recovering from wounds he had received during the

5   A contemporary engraving of the gunpowder riot at the Paris Arsenal,
6 August 1789.

storming of the Bastille: he had been attacked by revolutionaries who, having misidentified the National Guard uniform he was wearing, assumed he was the commandant of the fortress. Le Faucheux's son, although not directly involved, bravely joined his father in the midst of the angry mob.

The situation had become extremely threatening, with calls for immediate execution arising on all sides. It would not have been surprising had the directors fallen victim to mob violence at that point, yet Lavoisier continued to exhibit the sang-froid expected of a man of his breeding and station. Suddenly the crowd became convinced the true traitor was La Salle, who had authorized the shipment of powder. They surged into the city hall and fanned out down its corridors, determined to find him and hang him from the clock tower. Having been forewarned, La Salle had already made his escape. In subsequent letters to

271

the National Assembly he insisted that the order in question had originally been signed by the city police and that he was not going to
dishonor himself by covering up for incompetent functionaries.

> I have devoted my life to the cause of my country and the liberation of the
> Bastille has redeemed it, but my honor will not be compromised nor my
> life sacrificed to cover up for the stupidities of a police department that
> attempts to heap on me all the odium surrounding an order it issued.[2]

Bailly and his wife had been at Versailles that day and, when told
Lavoisier and his wife had been arrested, they hurried back to Paris.
Arriving at the city hall shortly before midnight, they found that
Lafayette had the situation well in hand. Lavoisier had already slipped
away while the crowd was searching for La Salle and had gone to an
apartment near the Palais Royal he maintained for meetings with his
scientific colleagues following sessions at the Academy.

Another story from this period focuses on symbols rather than mobs
and indicates how the revolutionary movement made use of pre-revolutionary forms of political and cultural contestation. The royal government had good reason to be nervous following the events of July and
early August, and the king's ministers were determined to avoid political confrontation whenever possible. They were well aware that all
forms of academic culture had become highly politicized during the
preceding decades.[3] In the realm of art, the biennial salons organized
by the Academy of Painting and held in the Louvre had been transformed into occasions for intense public debate. As everyone knew,
many of the published commentaries on these exhibitions, although
ostensibly about the works of art on display, were carefully coded
indictments of the academic establishment and the political culture that
sustained it.

The immensely talented Jacques Louis David contributed significantly to this politicization of art, first with his stridently moral *Oath of
the Horatii*, which was exhibited at the salon of 1785, and then with his
equally didactic *Death of Socrates*, exhibited in the salon of 1787.[4] These
works were greeted with ecstatic praise by David's fellow critics of the
Academy, who carefully explicated the paintings' political significance.
David himself thoroughly enjoyed applying his considerable skills as a
self-promoter and propagandist to undermining the legitimacy of aca-

demic standards. The next salon was scheduled to open in August 1789 and the head of the Academy of Painting, in an effort to avoid arousing further political discord in Paris, decided that none of David's recent paintings would be hung. But this attempt to reassert control over a public that had recently found its voice through commenting on academic exhibits did not go unopposed.

As early as 1787 David had decided to ignore the subject stipulated in the royal commission for his next major work and to concentrate instead on the painting that became *The Lictors returning to Brutus the Bodies of his Sons*. Thus for two years it had been widely if unofficially known that he intended to make this overtly political work his main contribution to the salon of 1789. It was also known that this painting, with its severe representation of republican honor, defiance of tyranny, and the defense of family against the sexual excesses of those who wield power, was both a narrative taken from classical history and a moral judgment on the recent sexual/political scandal surrounding the mesmerist Kornmann.[5] David had also intended to exhibit several recent portraits in the salon of 1789, including his painting of the Lavoisiers, but these too were excluded, evidently out of fear that pictures of figures prominent in the old regime might incite public agitation.[6] The public may have been unaware of or indifferent to the portraits, but they insisted on seeing the *Brutus*, and in the end they had their way. In David the Revolution had found a gifted and accomplished polemicist whose style and images would transform French art.

By July 1789 Lavoisier's career had reached its point of greatest expansion. His intellectual power, ambition, and devotion to public service remained undiminished, but with the onset of the Revolution circumstances inexorably began to constrict all his activities. Something similar might have occurred had he experienced failing health, family difficulties, or severe financial reversals, but in fact it was the Revolution that began to make his every task more difficult, the intended outcome of his projects more problematic. Lavoisier, of course, had no way of knowing how events would turn out and he continued to assume that the impediments thrown up by the Revolution would soon be removed by the establishment of a reformed political order. Such optimism, while reasonable, proved to be highly inaccurate as prediction. Yet we should not take the steady diminution of Lavoisier's career and influence from 1789 to 1794 as evidence that he was doomed by some tragic flaw in his personality or some fateful choice he had made in earlier years. His story is both simpler and more commonplace than that: he was squeezed by events that neither he nor anyone else could have anticipated or controlled.

Lavoisier constructed his public career as if it were a country estate; every few years he added a new wing to the core, but always in a way that maintained the balance and integrity of the whole. At the beginning of 1789 he and his wife occupied the entire structure, filling each of its rooms with the rich vitality of their own lives. But by the middle of that year there were signs that some of the wings would have to be closed off and their contents covered either for temporary storage or later disposal. The abrupt collapse of the Tax Farm eliminated a set of responsibilities that for two decades had occupied much of Lavoisier's time and had rewarded him richly as well. We have already seen how the government's reliance on *assignats* forced Lavoisier to comment on the nation's finances from positions of ever greater detachment. Political events as well soon carried the Revolution further and further away from the kind of limited monarchy that he believed should be at the heart of France's new constitution. One by one the subjects he was prepared to address were dragged beyond his reach by the unfolding drama of revolution.

How did Lavoisier manage his other areas of interest as the Revolution progressively dismantled his public life? What became of his science, and especially of his enormously successful and far from exhausted chemical research program? What became of the Academy of Sciences as the Revolution, in the name of liberty and equality, systematically pulled down the privileged corporations of the old regime? What role remained for Lavoisier in the Gunpowder Administration as France prepared to wage war against neighbors who dared not allow the French Revolution to succeed? In what other ways did Lavoisier participate in the Revolution before the last of the rooms in the estate that had been his career was converted into a prison cell? And finally, when stripped of all public responsibility and excluded from all meaningful public discourse, how did Lavoisier conduct himself as he awaited the end that his adversaries arranged for him?

When Lavoisier moved to the Arsenal in 1776, the center of his research activity shifted from the Academy of Sciences to the laboratory that adjoined the apartment he and his wife occupied. Although the Academy continued to provide the audience to which he reported his findings and the arena in which he sought the approbation of his peers, it was in his laboratory at the Arsenal that he now searched for novel facts and extended the range of his theories. This was not the work of a solitary alchemist muttering hermetic incantations over smoking beakers. Lavoisier's laboratory was a singularly well-equipped modern research center and he gathered around himself a talented and devoted group of younger investigators with whom he generously shared his time and thoughts. A recollection written by his widow, though scented with nostalgia, nicely conveys the intellectual pleasure he experienced while working in that laboratory with his colleagues.

> Each day Lavoisier sacrificed some time to the new affairs for which he was responsible. Science always had a large part of his day. He arose at six o'clock in the morning and worked at science until eight, and in the evening from seven until ten. One whole day a week was devoted to experiments. It was, Lavoisier used to say, his day of happiness. Certain enlightened friends, certain young men proud to be admitted to the honor of cooperating in his experiments, foregathered in the laboratory in the morning. There they breakfasted, there they discoursed, there they worked, there they performed the experiments that gave birth to the beautiful theory that has immortalized its author. Ah, it was there that a person needed to be to see and hear that man endowed with so fine a mind, so just a judgment, so pure a talent, so lofty a genius. It was by his conversation that it was possible to judge of the beauty of his character, the elevation of his thought, the severity of his moral principles. If ever any of the persons whom he admitted to intimacy can read these lines, I think the memory will not cross their consciousness without their being moved!
>
> It was into these sessions that the best workmen were admitted to make the machines that Lavoisier invented.[7]

Madame Lavoisier's mention of workmen and machines is commendable, for specially designed and precisely constructed scientific instruments played a central role in the research program carried on in the Arsenal. For philosophical reasons Lavoisier had from the very

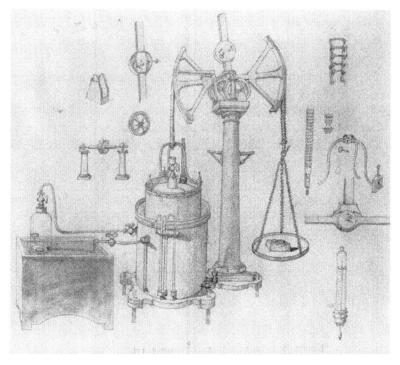

6   A preliminary pencil sketch by Madame Lavoisier for plate VIII of
Lavoisier's 1789 *Elementary Treatise on Chemistry*. The device pictured is a
gazomètre, an instrument that measures volumes of gases generated in
chemical experiments.

beginning of his scientific career focused on data provided by such
instruments as the barometer, the balance, and the thermometer. When
his personal wealth and circumstances enabled him to commission the
construction of new instruments, he quickly acquired experimental
equipment better than that of any other chemistry laboratory on the
continent. The balances, pneumatic chests for experiments with gases,
calorimeters, and vessels for analyzing and synthesizing water that
Lavoisier employed were built to his specifications by the prominent
instrument makers, Pierre Mégnié and Nicolas Fortin. The new apparatus was often designed by his associates, especially Meusnier and

Séguin, and Lavoisier gave them full credit for their contributions. Of course Lavoisier was not being novel in having instruments built and in equipping a private laboratory; his mentors Nollet and Rouelle, like many other eighteenth-century investigators and lecturers, had done as much. What distinguished his instruments, however, in addition to their great cost, was their innovative design and the precision of the quantitative data they made available. Many of these products of high craftsmanship also have a functional beauty that fortunately can still be admired in the exhibits at the Conservatoire des Arts et Métiers in Paris.[8]

Shortly before completing his *Elementary Treatise on Chemistry*, which was published in February 1789, Lavoisier began investigating certain topics in organic chemistry, the most notable being fermentation, the combustion of oils, and the analysis of vegetable and animal matter.[9] His interest in these topics was a natural extension of the analytic program that had led to the new theory of combustion, and he quickly succeeded in demonstrating that oxygen, hydrogen, and carbon are present in all vegetable matter. With this knowledge in hand, he returned to the problem of respiration, which from the outset had been as important as combustion itself in studies of the fixation and release of air.

While investigating respiration Lavoisier shifted his attention from the production of fixed air (carbon dioxide) to the consumption of oxygen. His recent studies on the combustion of oils, which he had shown consist of carbon and hydrogen, suggested that the oxygen consumed during respiration combines with both hydrogen and carbon to form water and fixed air. These reactions also produce heat, for according to Lavoisier's theory, when oxygen gas is fixed by carbon or hydrogen, it releases the "caloric," the substance that causes heat, with which it was combined in the gaseous state. Viewing the process of respiration in this way suggested a more comprehensive theory. Digestion could now be seen as a process that transforms foodstuffs into compounds of carbon and hydrogen that react with oxygen in respiration. Transpiration could be seen as a related process by which the body gets rid of the excess water and heat produced in respiration. Lavoisier therefore expected the body's rates of respiration, digestion, and transpiration to be closely related to its temperature and to the temperature of the environment in which it is located.[10]

In 1790 Lavoisier, with the assistance of Armand Séguin, began a series of experiments intended to illuminate these topics. Séguin designed an apparatus for capturing the vapors given off in respiration. These products of respiration were to be collected and analyzed while the factors governing the rate of respiration were systematically altered. The nature of these experiments, and of the laboratory in which they were carried out, are illustrated in two sketches by Madame Lavoisier. In the first, Séguin is at rest while breathing into a collecting apparatus; in the second he is exercising by working a pedal.[11]

While these sketches have significant shortcomings as technical illustrations, they are stylistically interesting. Madame Lavoisier, who is said to have studied with David, has depicted her husband in bold Davidian poses. In one sketch he is forcefully thrusting his hands into the basin holding the pneumatic collecting vessel; in the other he stands to the left, gesturing with both hands to the assistant carrying a sack of supplies toward the experimental apparatus. Madame Lavoisier has taken a compositional technique developed to celebrate political republicanism and applied it to the recording of experimental investigations in the republic of science. She used this novel artistic convention to assert graphically the high cultural status of science in the Enlightenment. Her sketches thus provide an illustration of the ways in which politically inspired modes of representation could be deployed when depicting other forms of cultural activity.

On 13 November 1790 Lavoisier delivered the first of a series of reports to public meetings of the Academy of Sciences on his and Séguin's respiration experiments.[12] He outlined the hypotheses they employed, but admitted frankly that much of what he supposed to be true had not yet been verified. Far from fully exploiting his enormous authority as an experimenter and theorist, Lavoisier openly acknowledged the tentativeness of his interpretations of the evidence he and Séguin had collected. As he told those attending this public meeting,

> perhaps I shall be obliged to make some modification in the doctrine that I have presented in this manner. I shall not hesitate to modify my opinions, even to reverse my steps, if new experiments force me to abandon the first course that I have followed.[13]

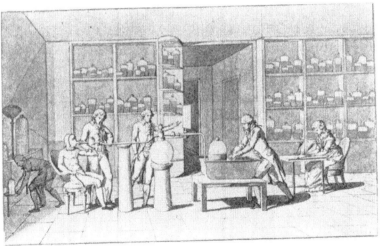

7   Two sketches by Madame Lavoisier, who is shown seated at the right, of the experiments on human respiration performed by Lavoisier and Séguin in 1790–1. The experiments measured the rates at which the subject, Séguin, consumed oxygen and gave off carbon dioxide when at rest and when working a foot pedal.

As Frederic Holmes has observed, these statements, made at a time when Lavoisier's theories commanded the "assent of almost every physicist and chemist in Europe," present their author in a most attractive light.[14] For Lavoisier the most profound pleasures associated with experimental enquiry clearly came, not from dominating the activities of his research associates or commanding the assent of his academic peers, but rather from pursuing truths about how nature operates. This is a point we should keep in mind when reflecting on the utilitarian arguments Lavoisier later felt compelled to deploy while struggling to shelter the Academy of Sciences from the effects of the revolutionary whirlwind.

In December 1791 Lavoisier became treasurer of the Academy of Sciences, and although he wished to pursue his research on respiration, he was soon obliged to turn it over entirely to Séguin so that he could devote his attention to urgent institutional problems. While there was no intrinsic reason the pursuit of science and the political revolution could not proceed independently, these two enterprises were in fact inseparably linked by history. Science, having basked in the legitimacy and hegemony associated with royal patronage, could not escape criticism of its former status when the Revolution called into question the modes of cultural authority sanctioned by the old regime.

Discussions of the authority of science naturally focused on the Academy of Sciences. The issues in dispute can be grouped into three general questions: (1) how should the Academy be reorganized so as to incorporate the new principles of liberty and equality; (2) who should decide which inventions are useful and how their authors should be rewarded; and (3) how is truth to be determined in a democratic society? In raising and addressing these questions about science and society, the Revolution opened the modern debate, which continues today, over the proper relations between science and politics.[15]

The Revolution was only a few months old and the Academy had just returned from its 1789 fall recess when the Duc de La Rochefoucauld d'Enville, an honorary member, challenged the Academy to reform its own constitution. The arguments he advanced at that November meeting were not simply a rehash of standard anti-academic polemics. He acknowledged that the Academy provided its members with a high degree of liberty and that to some extent it operated according to republican principles. What was now needed, he proclaimed, was a reorganization that would eliminate its remaining inequities and bring it into line with the political principles the National Assembly had adopted in August. The Academy needed to be reformed, La Rochefoucauld declared, not destroyed.

The issues raised in the debate that followed mirrored the larger issues being addressed in the National Assembly. Should the Academy have complete control over its own elections, or should the king, through his ministers, continue to have a voice in the selection of new members? It was widely agreed that the class of honorary members should be eliminated, but there were differences of opinion over how to remedy other forms of exclusiveness within the Academy. In the end it was decided that the only permissible class distinction would be between active members, meaning those who participated regularly, and ordinary members, who participated as amateurs. The assignment of members to sections concentrating on particular sciences was also de-emphasized in the name of equality and pensions were to be assigned strictly on the basis of seniority within the Academy as a whole. The debate between royalists and anti-royalists was especially intense. Those who, like Fourcroy, wished to free the Academy from all possibilities of ministerial manipulation insisted that the National Assembly alone be empowered to confirm elections; the royalists sought to preserve a role for the king. The Academy as a whole, guided by its constitutional committee, devoted months to hammering out a draft containing seventy-four proposals.[16] Although completed in September 1790, this new constitution was never ratified and the Academy, to its considerable embarrassment, continued to operate under regulations inherited from the old regime until its demise.

Many of those who wished to reform the Academy also sought to strip it of its traditional responsibility for evaluating practical innovations and deciding how they were to be exploited. The arguments advanced were both economic and political. As early as July 1790 the National Assembly was looking for ways to wrest control of innovation from the king's ministers. Members of the Assembly argued that instead of assigning social problems to salaried royal officials, rewards should be granted to those who made significant contributions to the welfare of the nation. The patent systems of Great Britain and the

United States provided models having several of the individualizing and liberalizing features these reformers hoped to adapt to French circumstances, and in January 1791 laws were passed recognizing the rights of inventors and establishing France's modern patent system.

Having deprived the Academy of one of its most important traditional duties, the Assembly was obliged to create new agencies for the alternative system it decreed. A Patent Office (Bureau des Brevets et Inventions) was established to register and provide information on new inventions, and in September 1791 an Advisory Office on Practical Arts and Trades (Bureau de Consultation des Arts et Métiers) was formed to evaluate proposed innovations and determine which were to receive rewards. The members of the Academy of Sciences were well aware of the political implications of these moves. They resisted being pushed aside by critics who claimed that in the past their evaluations of technical proposals had been ill-informed and high-handed. Quietly but effectively they positioned themselves to gain control of the new institution to which their former responsibility for mediating between science and practice had been assigned. Although the Advisory Office was supposed to consist of an equal number of members from the Academy and from artisanal groups, in fact representatives from the Academy of Sciences and other scientific societies captured a commanding majority.[17] For two years the academicians demonstrated their political acumen by holding on to the substance of power after having been deprived of the institutional authority they had enjoyed in the old regime.

During those two years Lavoisier was one of the leading members of the Advisory Office.[18] He wrote numerous reports on topics ranging from a proposal for a mineralogical atlas of France, an old favorite of his, to the technical problems involved in making the paper, designs, and dyes used in the printing of *assignats*. He was also involved in drafting and revising a comprehensive proposal for a national system of education that the Advisory Office submitted to the Convention in September 1793.

The Advisory Office's education proposals were founded on a plan that Condorcet had presented to the Legislative Assembly in April 1792.[19] Had Cordorcet's plan been adopted, the government would have assigned responsibility for supervising all levels of education in

France to an autonomous National Academy of Sciences and Arts. The new system was to include a significant commitment to technical education, and technology was to be a major concern of the new national academy.

This conjoining of a new national academy and a plan for national education, the latter being a subject the revolutionaries considered especially important, reveals that the leading figures in the Advisory Office were pursuing a carefully constructed strategy. The new national academy would possess much greater autonomy and exercise more comprehensive authority than the now much maligned Academy of Sciences had ever enjoyed and it would provide support for all the activities that had previously been sustained by its predecessor. This Turgotist vision of a single academy charged with promoting science throughout the nation and making science the dominant mode of culture within France, while not unopposed, did eventually triumph: in 1795, the Convention, just before being dissolved, agreed to transfer the functions of the Advisory Office to an Institut National des Sciences et des Arts.[20] By that time, of course, Lavoisier and Condorcet were no longer alive. Had they been, they would have welcomed, as did their surviving colleagues, the success of their brilliant strategy for extending, during a period of revolutionary upheaval, the hegemony of science in French culture.[21]

The question of how truth is to be determined in a democratic society raised both institutional and cognitive problems. The revolutionaries, many of whom had experienced rejection during the old regime, vigorously attacked state-supported corporations, which they claimed promoted aristocratic science. They championed instead voluntary societies in which, they believed, democratic science would flourish. Early in 1790 the chemist, editor and ferociously pro-phlogistonist Lamétherie complained that the salaried academicians in Paris monopolized all the nation's scientific posts and excluded worthy scientists from the Academy. Such exclusiveness, while permissible in private organizations, is insupportable when indulged in by servants of the state.

In voluntary associations, such as the Royal Society of London, no injustice is committed in rejecting an otherwise qualified candidate, since the association – being a purely voluntary one – has the right to reject anyone whose moral qualities do not correspond to those shared by the majority of members, as is the case in political clubs; but such is not the case with associations salaried by the nation.[22]

The Lycée, a private institution that offered public instruction primarily in the sciences, provided one model for the kind of free society Lamétherie had in mind.[23] Originally called the Musée when opened by Jean François Pilâtre de Rozier in December 1781, this private academy counted among its founders Condorcet, Fourcroy, and Vicq d'Azyr, but not Lavoisier. In 1785 Pilâtre de Rozier, an avid balloonist, had the misfortune to become the first aeronautical fatality, and following his death the Musée was reorganized as the Lycée. Shortly thereafter Fourcroy, who had recently converted to the antiphlogistic chemistry, was appointed professor of chemistry and natural history in the Lycée. His courses were well-attended, but the beginning of the political Revolution once again placed the Lycée in peril. When it was reorganized a second time in 1790, Lavoisier and La Rochefoucauld of the Academy of Sciences joined and Fourcroy was elected president.

As the polite audience for public education melted away in the heat of revolution, the Lycée found it increasingly difficult to carry on. In 1792 its directors therefore decided to seek financial support from the government. A subsidy from a fund for the encouragement of the arts and sciences was arranged, but it was accompanied by a warning. The Lycée was cautioned that its lecturers must not make "imprudent speeches" or advocate unpatriotic "deviations." This warning was repeated even more sternly a year later when a second request for financial support was submitted to the Convention's Committee on Public Instruction, on which Fourcroy sat as a member. In November 1793, several months after all the royal academies had been suppressed, Fourcroy reported to the Lycée that no subsidy would be granted unless it severed its connections with shareholders who were considered counter-revolutionaries. A committee was formed to carry out the unwelcome task of purification, which certainly made a mockery of the notion that the Lycée enjoyed greater freedom than had the Academy; it submitted its report a few days after Lavoisier was arrested for having been a tax farmer. The Lycée was immediately renamed the

Republican Lycée; its furnishings and decorations, which were tainted with feudalism, were discarded and replaced with symbols of liberty and equality. Only twenty-six of the shareholders survived the committee's scrutiny; the other sixty, including Lavoisier, were excluded.

Prior to his arrest in November 1793 and his exclusion from the Lycée, Lavoisier played an active role in a different but related association, the recently organized Lycée des Arts. The leaders of this consortium of educational foundations were full of enthusiasm and Lavoisier, although reluctant to take on yet another administrative assignment, agreed to serve on its governing board as both the Lycée's representative and, for two months, its president.[24] A report he prepared on the Lycée des Arts was read to a meeting of the Lycée in June 1793.[25] Despite his concerns about the financial condition of the Lycée des Arts, Lavoisier thought it might succeed in transforming itself into a central society for science and the arts. It was hoped the association would provide a meeting place for representatives of many different learned societies and offer a broad range of courses.

Lavoisier was clearly attempting to transform an organization that the more radical revolutionaries considered free and egalitarian into a comprehensive institution capable of taking on some of the cultural duties he and his Turgotist colleagues believed ought to be supported and supervised by state agencies. His goals for the Lycée were the same as those he pursued as a member of the Advisory Office. As the prospects for the Academy of Sciences faded, Lavoisier was doing all he could to ensure that a cultural institution capable of carrying on the Academy's work would emerge from the ashes. The aggressive strategy he and his colleagues followed was a manifestation of their self-assurance, confidence, and hope. By the end of 1793 the Revolution had done great damage to the old regime's scientific institutions, but it had not directly attacked the assumptions that had guided them concerning such questions as what science is and how it should serve the nation.

Breaking the spirit of science was certainly not one of the professed goals of the revolutionaries. They set themselves the task of cleansing science by separating those parts of it that were of enduring worth – its facts, methods, theories, and above all its purported utility – from its

285

insupportable historical excrescences – its hauteur, elitism, idleness, and aristocratic exclusiveness. Although no prominent revolutionaries condemned science as a way of obtaining reliable knowledge about nature, many of them, like other analysts of science throughout history, proceeded on the appealing but unwarranted assumption that the cognitive aspects of science can be neatly separated from its institutional support structure. They wanted a science that was free, egalitarian, and patriotic, but they found that the only way they could work toward this goal was to reconstruct the institutions of science. It was not a lack of knowledge or ability that prevented them from articulating a novel cognitive account of science, one that would ensure the political correctness of all future scientific theories. They were blocked, rather, by the intractable and irreducible complexity of the problems they addressed. Science, like politics, turned out to be much more complicated and less easily manipulated than the radicals anticipated.

There is a fundamental tension between democracy and science that cannot be eliminated by political willfulness or clever redefinition. Having committed themselves to certain political principles, the Jacobins made a serious attempt to impose their social and political views on science. Angered by the hierarchical organization of science in the old regime, they longed for a knowledge of nature that would be equally accessible to all men. And when faced with proposals that attached more importance to truth than virtue, they struggled to maintain what they considered the correct balance between science and politics. As a critic of Condorcet's education plan complained, "to be happy the people of France need no more science than is required to be virtuous . . . Rome was happiest and most flourishing when it was entirely agricultural."[26] While moving nervously back and forth between political principles and descriptive statements about nature, revolutionary critics often expressed themselves in ways that appear to call for a radical reconceptualization of science itself. But such statements should not be taken out of context, for even the most radical revolutionary spokesmen, like their Marxist descendants, were seeking to purify science, not transcend it.

Even Bernardin de Saint-Pierre, an especially visionary critic of science, insisted that "it is not science itself that I am blaming."[27] What Bernardin objected to was the priestly arrogance of scientists, and he articulated his objection by writing a fable in which every man is equally capable of attaining truth about nature. Roger Hahn has summarized his beliefs as follows:

> Truth is not the property of a titled, intellectual elite, possessing special qualification for grasping it. To be understood, nature must be approached in a direct, unaffected manner, with a heart filled with simplicity and virtue. Every man has access to nature, and, in fact, the lower classes are more likely to comprehend its deepest meaning because they habitually commune with it. In the search for truth, erudition is unnecessary and even detrimental.[28]

In Bernardin we encounter the unmistakable voice of romantic naturalism, which was soon to inform and become identified with the European apprehension of nature and the sublime, yet it took some time to transform this affective reaction against aristocratic science into a well-articulated alternative theory of knowledge. The French Revolution, meanwhile, concentrated on eliminating aristocratic political institutions. It set out to cure the ills of science with democracy rather than with subjectivism or occultism. When Bernardin returned from the land of fables to the world of political action, his recommendations for science once again addressed the institutional rather than the cognitive structure of knowledge.

> If the sciences and letters have an influence on the prosperity of a nation – which we cannot doubt – it might be proper to have the nation elect the members of its academies, as they do members of other assemblies. New knowledge must be held in common, as are the other riches of the state. When the academies elect their own members, they become aristocracies harmful to the republic of science and letters. Since one can only be admitted by courting their leaders, one must tie himself to their systems of thinking. Errors are thus maintained by the authority of corporations, while isolated truth finds no partisans.[29]

While Lavoisier was doing what he could to turn the Advisory Office, the Lycée, and the Lycée des Arts toward the preservation of science, he was also looking for ways to shield the Academy of Sciences from the Convention's root-and-branch attack on the old regime. The primary strategy pursued by Lavoisier, the Academy's treasurer, and Condorcet, its secretary, was to stress the Academy's contributions to the establishment of a national system of weights and measures, an initiative of great concern to the Revolution. As criticism of all the academies intensified, they concentrated on saving the Academy of Sciences alone, which they insisted had been less tainted by royal patronage than the other academies and was infinitely more useful to the nation. It was a rear-guard action, and rather ragged at that, but Lavoisier simply could not stand by idly while the institution within which he had made his scientific career was dismantled. Although he was aware that continued public resistance might at any moment cost him his life, Lavoisier defended the Academy with a boldness that bordered on folly.[30]

Creating a single system of weights and measures was a reform that appealed to all liberals. Many of the booklets of instructions for delegates to the Estates General called for this reform and the members of the National Assembly finally turned to it near the end of their first year in office. In May 1790 Talleyrand, responding to a similar initiative in England, proposed that responsibility for creating a universal system of weights and measures be assigned to the Academy of Sciences and that the Royal Society of London be approached on the possibility of forming an international commission.[31]

In France an elaborate research effort was promptly organized within the Academy of Sciences. A central committee consisting of Borda, Lagrange, Laplace, Monge, and Condorcet was formed to plan and coordinate the work. In March 1791 this committee recommended that the basic unit of length be related to a fundamental dimension of the Earth derived from a carefully surveyed arc of latitude. It also created five additional committees, each of which was charged with determining one of the fundamental units of length or weight to be used in constructing the new system. Lavoisier and Haüy were to determine how much a given volume of distilled water at zero degrees would weigh in a vacuum. In September the National Assembly approved this plan and appropriated 300,000 livres to support its implementation. Lavoisier was appointed treasurer for the central committee and also served in effect as its secretary.

For a few happy months the members of the various committees on weights and measures had the pleasure of pursuing research that was technically sweet, well financed, and of indisputable national importance. New instruments were designed and ordered, surveying expeditions were organized and dispatched, and interesting new problems and challenges were identified. While other members of the Academy were agonizing over constitutional reform and defending themselves against anti-royalists and angry artisans, the reformers of weights and measures were demonstrating that a state-supported scientific society could organize its members into a practical task force. The questions addressed were ones that both scientists and legislators found worthy, if for somewhat different reasons. Whether the approbation accorded to this manifestation of practical science would be enough to save the Academy as a whole remained to be seen.

The Academy's future was looking increasingly grim. Falling attendance, disorderly accounts, and the government's growing reluctance to provide funds all made Lavoisier's duties as treasurer burdensome.[32] Yet he continued to plead the Academy's case while also supporting his colleagues with personal loans, one of which went to the chemist Sage, a vigorous opponent of the new chemistry.[33] In July 1792, shortly after agreement had been reached on the nomenclature for the new system of weights and measures, Monge retired from the central committee and Lavoisier replaced him. In November the committee reported to the Convention on their progress. They indicated that the entire project would be completed early in 1794 and openly credited the Academy with the success of their effort. Henri Grégoire, the constitutional Bishop of Blois, replied for the Convention and praised the members of the committee for their work, but he pointedly avoided mentioning the Academy.

The following month the Convention took up the education proposals Condorcet had submitted the preceding April; the debate that ensued revealed the depth of antagonism toward academies.[34] Condorcet had proposed that a National Society, a sort of super-academy, be created to supervise a national system of education. But by the time these proposals were considered by the Convention, the Jacobins had turned against Condorcet and the debate quickly became personal and nasty. Brissot defended Condorcet and urged the members of the Convention to remember that their colleague had devoted himself to the cause of the people for many years.

Can you, like him, point to thirty years of campaigning with Voltaire and D'Alembert against the throne, superstition, ministerial and parliamentary fanaticism? . . . You tear Condorcet to shreds even though his revolutionary life is one series of sacrifices for the people: *philosophe*, he made himself a politician; academician, he made himself a journalist; noble, he made himself a Jacobin.

Robespierre replied with a condemnation that was equally sweeping:

If our leaders of liberty are academicians, friends of D'Alembert, I have nothing to reply save that reputations in this new regime cannot be based upon reputations in the old; that if D'Alembert and his friends ridiculed the priesthood, they also befriended kings and the powerful.

As criticism of the academies mounted, it became apparent that some action would soon have to be taken. In March 1793 the busts that decorated the hall of the Academy of Sciences were removed; in July the hall was stripped of all signs of royalty.[35] Lavoisier, working closely with Joseph Lakanal, a member of the Convention and of its Committee on Public Instruction, struggled feverishly to separate the Academy of Sciences from the other academies, his hope being that the Convention could be persuaded by the usefulness of science that his academy should be spared.[36] A memorandum he drafted in July at Lakanal's request begins with the following assertions:

The Academy of Sciences should be evaluated in two different ways: first as a company of many men of learning who work in common for the advancement of science, for the progress of the arts and national industry, and for the stability of the human spirit; second, as a permanent and continuously active committee that the duly constituted authorities can consult and make use of for such purposes as require their attention.[37]

This is clearly the rhetoric of public justification rather than a dispassionate and balanced statement of Lavoisier's true beliefs concerning the Academy's mission. The same is true of his claim that the republic of letters and the republic of science are fundamentally different.

The sciences are not like other literary pursuits. The man of letters finds everything he needs to develop his talent in society . . . It is not the same in the sciences, most of which cannot be successfully pursued by isolated individuals. A collective effort is required; often, to come to a conclusion, contributions from many different specialists are needed . . . All the sciences help one another in constructing the great edifice of human knowledge.[38]

Lavoisier and Lakanal were contending with impassioned Jacobins eager to root out the last vestiges of a regime they execrated; their tactics were understandably opportunistic rather than disinterested. If the Academy of Sciences could only be saved by wrapping it in the mantle of revolution, Lavoisier was prepared to do so.

Finally, the Academy of Sciences is at this very moment charged by the legislative body with duties of the highest importance . . . We ask the Convention to pause for a moment to consider one of the finest enterprises undertaken to achieve human happiness, one of the greatest legacies of the French Revolution, the establishment of a universal system of measurement . . .

This plan can only be brought to conclusion by the Academy of Sciences, which alone can carry it through. Consequently, the completion of this task, which is of interest to everyone on the earth, is closely tied to the existence and preservation of the Academy.

Does the National Convention wish to terminate the progressive development of the arts and sciences in the French Republic? Does it wish to suspend investigations that it has itself initiated?[39]

This unlikely defense almost succeeded. Early in August Grégoire, when speaking for the Committee on Public Instruction, introduced a bill to the Convention that included most of the arguments Lavoisier had formulated, and it appeared that while the other academies, including the equally useful Société Royale de Médecine, would indeed be suppressed, an exception would be made for the Academy of Sciences. But in the end categorical condemnation triumphed. Marat, the implacable critic of academies, had been murdered a few weeks before this debate and his cause was now championed by the painter David, himself a member of the Committee on Public Instruction. David delivered an impassioned speech to the Convention, excoriating the academic establishment, denouncing the Committee's recommendation, and demanding that all academies, without exception, be abolished. He carried the day and on 8 August 1793 the Academy of Sciences and its sister academies formally ceased to exist.[40]

Lavoisier was not inclined to accept defeat. He promptly proposed that the proscribed Academy of Sciences be reconstituted as a Free and Fraternal Society for the Advancement of the Sciences; he also petitioned the Committee on Public Instruction for continued support for the Academy's ongoing projects. Lakanal slipped a bill through the Convention instructing the scientists to continue their work and promising financial support. This promise could not be fulfilled, however, for other bills had been passed authorizing governmental seizure of the Academy's property. When, under Lavoisier's leadership, the members of the Free Society attempted to hold their first meeting in the Academy's rooms, they found the doors bolted and sealed. The will of the Convention was not to be thwarted; the Academy's long life was irrevocably at an end.[41]

The suppression of the Academy eliminated the forum to which Lavoisier had reported his discoveries and presented his theories; his removal from the Gunpowder Administration on 15 August 1792 had previously denied him access to the laboratory in which he had conducted much of his research. Although the Gunpowder Administration had served the nation admirably, it had not been able to shed the stigma attached to all offices established in the old regime. As early as January 1791 Lavoisier and his fellow directors were forced to defend themselves against charges brought by leaders of the Revolution.[42] Nine months later Lavoisier was dismissed and the number of directors was reduced to three, but when he protested, he was allowed to keep his apartment in the Arsenal.

In February 1792 Lavoisier was reappointed as a director, but late in the following summer he was forced to resign and vacate his apartment and laboratory. No longer able to pursue his research program in physiological chemistry, he turned to projects that did not require access to a well-equipped laboratory. By the end of the year he was hard at work on an encyclopedia of chemistry and, like Cordorcet, Dupont, and several other scientists driven from public life by the Terror, he passed his hours of enforced retreat recording his reflections and arranging his papers for posterity. His career as a creative chemist was over even though his capacity for creative work remained undiminished.[43]

The Academy of Sciences was eliminated just as the revolution was entering its most extreme phase. The revolutionary conception of republican governance, a concept of enormous symbolic power in French political culture, was again being redefined.[44] So long as the absolutism of the old regime had endured, it was understood that only members of certain social and cultural communities, such as those included in the republics of letters and sciences, could enjoy the liberty and equality that come with citizenship. The collapse of the old regime encouraged many to believe the way had been opened for universal political republicanism in France. The aristocratic and bourgeois liberals who directed the revolutionary impulse during its first two years set about transforming France into a constitutional monarchy. They were unable to maintain control, however, and the execution of the king in January 1793 extinguished what little hope remained that the Revolution might be brought to a close with the adoption of a moderate form of government. As the consequences of disestablishing the monarchy unfolded, the politics of will became predominant. As France hovered on the brink of political chaos, the new leaders of the Revolution grasped at ever more coercive means for maintaining public order. In 1793 they declared that henceforth France would be a republic of virtue; the instrument they chose to ensure the ascendancy of virtue was terror.

The policy of terror was embraced in a desperate act of acquiescence. On the morning of 5 September the Convention was invaded by a mob of sans-culottes demanding bread and the guillotine. Terror would provide bread, they believed, by eliminating the conspirators whose schemes lay at the root of all the dangers and hardships besetting the nation. Thus, while the Committee on Public Safety authorized the Terror, it did not concoct this political strategy on its own. The demand for terror was, as François Furet has pointed out, intrinsic to certain political convictions that form "a characteristic feature of the mentality of revolutionary activism."[45] The Convention responded to this mentality on 17 September by passing the infamous Law of Suspects, legalizing the imprisonment of anyone whose loyalty to the republic and revolution had been questioned. The distinction that separated political opposition from treason and criminality was eliminated at a stroke. In the same month the Revolutionary Tribunal in Paris was reorganized and made more efficient; as a result the procedural protection formerly provided by the judicial system was rendered ineffective. Those accused could still be acquitted, but as the fury mounted acquittals were handed down far less frequently.

Marat articulated the politics of terror for a revolutionary leadership that found the doctrines he espoused increasingly unavoidable. For years he had been lashing his enemies with virulent accusations, and when terror became official policy, his voice helped focus its demonic energy. Those whom he had attacked with little effect in earlier years now had good reason to fear him. Although murdered by Charlotte Corday in June 1793, Marat remained the pre-eminent apostle of terror. His martyrdom appeared to validate his repeated assertion that all those who opposed the people he represented were part of a massive conspiracy against the Revolution.

The turn to terror was forced, if not justified, by events. February 1793 opened with the declaration of war against Great Britain and the Dutch Republic; it closed with food riots in Paris. March saw the creation of the Revolutionary Tribunal in Paris and the formation of the Revolutionary Armies. In April 1793 the Convention created the Committee on Public Safety to invigorate its executive powers; Maximilien Robespierre joined the Committee at the end of July. Early in August the academies were suppressed; on 5 September the Committee on Public Safety took effective control of the government, began mobilizing the nation for war, and declared terror to be the political order of the day.

The suppression of the academies was thus just one in a series of acts that marked the emergence of a severe new mode of political discourse. The discourse of virtue, as structured by the reign of terror, proved to be extraordinarily threatening to those who had formulated and carried out public policies in the old regime and early years of the revolution. No one could predict how long this political storm would last or where its destructive bolts would next touch down, but there was no mistaking the threat it posed to all those who participated in public life.

The Terror succeeded in running Lavoisier to ground, but it only did so after radically subverting traditional procedures for objectively determining whether public officials had properly carried out their duties. The disjunction between civic life in a society governed by laws and public life in the republic of virtue was absolute. The Lavoisiers were not unfamiliar with the suspicion and anger that poisoned the hearts of many of their fellow countrymen. Years earlier, on 5 October 1789, Gouverneur Morris, who later served as the American Minister to

France, stopped by the Arsenal to visit the Lavoisiers. He found that Madame Lavoisier had been detained in town by a mob of women who had stopped her carriage and forced her to descend and march along with them; this was the crowd that continued to Versailles and brought the king and his family back to Paris.[46] Lavoisier had also repeatedly encountered the Revolution in arms. On 10 August 1792, the bloodiest day of the Revolution so far, mobs of communards in Paris stormed the Tuileries, seized the king and his family, and brought the long era of monarchy to an end; Lavoisier spent the day commanding the soldiers guarding the powder magazine at the Arsenal. Less than a year later, on 31 May 1793, he stood guard at the already bloody Place de la Revolution, where the following year later he would be guillotined.[47] The Lavoisiers obviously hoped to survive the Terror, but they were not ignorant of the ferocious mentality that drove it onward.

Those who directed the reign of Terror, being little concerned with science, were not prepared to overlook charges of latent royalism among those who had distinguished themselves in the study of nature. Lavoisier was not imprisoned, tried, and executed because he was a scientist. Although he did ask that his life be spared so that he could continue to serve the state through science, this personal plea was no more effective than similar pleas had been in defense of the Academy of Sciences. But Lavoisier's scientific colleagues, unlike the administrators of the Terror, did consider his extraordinary record of achievement germane to the determination of his fate, and they did not desert him. Following the imprisonment of the tax farmers in late November 1793, letters, petitions, and deputations in his behalf were received from the Advisory Office on Practical Arts and Trades, which continued to list Lavoisier as its president; from the Committee on Weights and Measures, which the Convention had established as an independent body following the suppression of the Academy; from the Lycée des Arts; from the Committee on Assignats and Coins; from the Gunpowder Administration; and from colleagues such as Cadet and Baumé, the latter of whom remained an unreconstructed phlogistonist.[48] Guyton de Morveau was on his way to Belgium to join the army when the tax farmers were tried; it is unlikely he could have intervened in any case, for, as he later learned, his name was on a list of proscribed individuals drawn up by Robespierre. When Fourcroy heard that the tax farmers were to be arraigned before the Revolutionary Tribunal, he boldly pleaded for Lavoisier's life before the Committee on Public Safety, of which he was not a member. Robespierre heard him out without comment and Fourcroy, whose own life was in danger, was unable to prevent what had become inevitable.[49]

The central issue was not the way Lavoisier had performed his duties as a tax farmer and director of the Gunpowder Administration. A full accounting of the General Farms and the Company of Farmers General had been called for early in the Revolution, but it had not been completed by the time the tax farmers were imprisoned and tried. Although Lavoisier repeatedly refuted charges that the tax farmers and directors of the Gunpowder Administration had violated the public trust, the aura of suspicion and conspiracy that dominated the Terror deprived his evidence and arguments of weight. Two centuries of additional investigation have still not turned up any evidence that Lavoisier was guilty of misconduct in the discharge of his many public duties. But again we must note the irrelevance of this absence of evidence to the internal dynamics of the Terror. The Terror was a collective response to extreme circumstances; it was guided by the Committee on Public Safety's conviction that the nation had to be ruthlessly purged so that its forces could be marshalled for the defense of the fatherland. Reason and methods for reliably connecting interpretation to reality counted for little in a world preoccupied with fear, anxiety, and a desire for retribution.

The story of Lavoisier's last months can be quickly sketched.[50] The Convention was growing angry over the delay in dissolving the Company of Tax Farmers. When Antoine Dupin, one of its members who had been an administrator in the tax farm, suggested in September 1793 that he and a group of other former employees could settle the Company's accounts in short order, the Convention happily turned the problem over to them. More concerned with avenging personal grievances than rendering justice, these examiners quickly set about drawing up the charges expected of them. By the end of November Lavoisier, his father-in-law Jacques Paulze, and other members of the Company had been arrested. They continued to defend themselves as vigorously as was possible under the circumstances, but they knew they were not likely to prevail.

How did Lavoisier and his wife respond to being deprived of the social approbation and support to which they were accustomed? By the time Lavoisier was imprisoned, the opportunity to flee France for the safety that lay beyond her borders had passed.[51] They were therefore obliged to confront as individuals the republic of virtue in all its terrifying immediacy. Adversity laid bare their personalities and tested their characters. The significance of all they had achieved was devalued by a radical shift in the standards of public behavior. If they wished to draw strength from the doctrines and beliefs they had lived by, they could now do so only in private. Isolated and disgraced, they did what they could to preserve their sanity while struggling to survive in a world turned upside down.

The political culture of the republic of virtue was personal and emotional, and in that sense Romantic. Danton and Robespierre claimed to represent the spirit of the French people rather than their interests; they controlled the legislative arena by exerting their political wills and imposing their moral passions on others rather than by relying on reason and evidence. The rhetoric of politics had shifted from functionalism to personal belief, a shift that eliminated the distinction between public and private just as thoroughly as the turn to terror had eliminated the distinction between political opposition and treason. While Lavoisier may have understood abstractly how profoundly France's political culture had changed, he persisted in defending the utilitarian view of public service to the end. Like others who had made successful careers in the old regime, he distinguished sharply between his private life, which he continued to conduct with a dignified self-control, and his public career, in which he was guided by his well-known commitment to the common good. Although he never indulged in sweeping condemnations of the revolution as a whole, Lavoisier went to his death opposing its abandonment of the Enlightenment he had known.

In an autobiographical note drafted during his months in prison Lavoisier proudly defended his career as a scientist and citizen.[52] Unfazed by the prospect of the guillotine, he directed attention to those acts and achievements that he thought would earn him credit in the republic of virtue. He ignored completely his years of work on the Tax Farm and the Gunpowder Administration. He stressed instead his

8 An engraving of a sketch of Lavoisier made when he was imprisoned in 1793–4

international fame as a scientist, his many "discoveries of importance for the arts, the sciences, and humanity," and the fact that he devoted part of his fortune to supporting these activities. No mention was made of his commitment to the centralizing and liberalizing doctrines of the physiocrats, but considerable attention was given to his efforts to improve agricultural productivity. Lavoisier insisted with some justice that "well before the revolution had begun he had demonstrated his

298

commitment to the principles of liberty and equality." He supported this claim by citing the civic donations he made during periods of grain shortage. He also listed the many ways he had served the Revolution as an elected representative, as a financial adviser, and as a member of the Paris guard.

This apologia inevitably evokes mixed feelings. Lavoisier evidently believed he could still redeem his public reputation by shrewdly re-shaping the public's perception of his career. His unwavering commit-ment to the public good and his indomitable spirit command our respect, even though we can see, as he apparently could not, that the rhetorical tactics he was employing were bound to be unavailing in the political culture of the Terror. Lavoisier remained true to the end to his vision of himself as a honorable and effective public man, even though the prevailing standards of public conduct had been radically altered.

Lavoisier's private appraisals of his situation were less rhetorical and reveal the personal agony he hid behind a stoic mask of civic fortitude. Shortly after his arrest he wrote to his wife following one of her visits. The inadequacy of words and the futility of hope hang like a pall over the forced cheerfulness with which he sought to alleviate her sorrow.

> My dearest friend, you are beset by much pain and weariness of body and soul that I cannot share. Be careful not to sacrifice your health, for that would be the greatest misfortune. My career is well advanced and I have always enjoyed a happy life. You have made it so and continue to do so by all the signs of affection you show me. When I am gone I will be remembered with respect. My work is done, but you, who have reason to hope for a long life, must not waste it. Yesterday you appeared to be sad. Why should you be, when I accept what will happen with resignation and will consider anything that is not lost as a victory. There is also reason to hope we will be together again. Until then, your visits provide me with moments of happiness.[53]

The day before his execution Lavoisier wrote to his cousin Augez de Villers, whose brother, Augez de La Voye, had written to Lavoisier twenty-four years earlier to congratulate him on his election to the Academy of Sciences. Lavoisier hoped to ease the grief of those who would mourn his death, and yet his letter also reveals his anger and dismay at the disjunction between the life he led and the death that awaited him. Having been denied justice as a citizen of France,

Lavoisier was reduced to conducting himself with dignity and consideration for others as he contemplated his unintelligible fate.

> I have enjoyed a reasonably long and above all a happy life and I trust my passing will be remember'ed with some regret and perhaps some honor. What more could I ask for? I will probably be spared the troubles of old age by the events in which I find myself embroiled. I shall die while in my prime, which I count as another of the advantages I have enjoyed. My only regret is not having done more for my family. I am sorry to have been stripped of everything and to be unable to give you and others tokens of affection and remembrance.
>
> Evidently it is true that living according to the highest standards of society, rendering important services to one's country, and devoting one's life to the advancement of the arts and human knowledge is not enough to preserve one from evil consequences and dying like a criminal!
>
> I am writing today because tomorrow I may not be allowed to do so and because I find it a comfort in these final moments to think of you and others who are dear to me. Be sure to tell those who are concerned about me that this letter is addressed to them all. It is probably the last I shall write.[54]

Lavoisier went to his death with his spirit unbroken; his wife, although her life was spared, was more traumatized psychologically by the experience of the Terror than he had been. Whereas he sustained his sense of worth by recalling his distinguished public career, she had only her private life to fall back on, and this the Revolution had been systematically destroying.[55] Madame Lavoisier was only thirty-five years old when her father and husband were arrested. During their years at the Arsenal, she had happily served as hostess to the circle of scientific collaborators that gathered around her husband and as a valuable research associate in the laboratory, but all this came to an end with their eviction in August 1792. After that she had been forced to adjust to constantly diminished and increasingly embattled circumstances.

Having lost her mother while a child and having no sisters or children, Madame Lavoisier had devoted her considerable energy and intelligence to supporting her husband's work by entertaining his colleagues and developing her own talents as a laboratory assistant, artist, musician, and translator. She had, in short, mastered many of the skills required for a public life in the polite culture of the old regime, but she had always deployed them within the confines of her private life as an affluent Parisian wife. In June 1789 Gouverneur Morris, who had rather traditional ideas as to what women were good for, noted in his diary after dining with the Lavoisiers that "Madame appears to be an agreable Woman. She is tolerably handsome but from her Manner it would seem that she thinks her forte is the Understanding rather than the Person."[56] When in the Terror Madame Lavoisier found herself without friends, purpose, comfort, or status, she was evidently left feeling extremely isolated and vulnerable. Her husband's all-consuming commitment to preserving the institutions of French science did little to relieve her loneliness. She had long known there was a streak of malevolence in the Revolution; following the arrest of her husband and father she had to face its consequences on her own.

It was in these circumstances that Madame Lavoisier turned for comfort to an old family friend, the long widowed Pierre Samuel Dupont. Their surviving letters contain shards of evidence that indicate they had a brief, intense, and profoundly unhappy love affair.[57] Had this affair been pursued in the heyday of the old regime, we might ask what it tells us about the codes of conduct that prevailed in high bourgeois society during the Enlightenment. But the threatening situation the Lavoisiers and the Duponts found themselves in during the closing months of 1793 suggests this *mésalliance* should be viewed primarily as evidence of how thoroughly Madame Lavoisier had been unsettled by the maelstrom of the Terror. The unforced tenderness of the letter Lavoisier wrote her from prison and the fact that after the Terror had passed she chose to devote the remainder of her life to vindicating his name strongly indicate that it was revolutionary violence, rather than a loss of loyalty or affection for her husband, that led her to cling briefly to Dupont. Seen in this light, an episode that otherwise might be allowed to pass without comment takes on the aura of healthy human passion in a setting of unremitting moral squalor. Although hardly a representative event in the lives of the Lavoisiers, Madame Lavoisier's affair with Dupont can thus be seen as representing a response to social evil that in the nineteenth century would be glorified in numerous romantic operas.

The Convention's relentless pursuit of the former tax farmers visited ever greater humiliations on Madame Lavoisier. When the tax farmers were arrested their property was seized and placed under seal. In her desperation she appealed directly to Dupin, who was coordinating the attack on the tax farmers, begging him to use his influence to save her husband, but to no avail.[58] In June, following the tax farmers' execution, she and the other heirs were arrested and jailed; two months later she wrote abjectly to her local revolutionary committee to declare her attachment to republican principles. By that time, however, Robespierre had fallen and the grip of the Terror was beginning to loosen; in August she was released from prison. The Convention was not finished with the tax farmers, however, and in September she found herself utterly without means and dependent on the charity of a former servant. By the end of the year some of her property had been restored and the worst was passed. Although she survived until 1836, it is no exaggeration to say that she never fully recovered from the grievous spiritual wounds she suffered during the Terror.

For Lavoisier the end came swiftly. On 5 May 1794 Dupin presented his report on the Tax Farm to the Convention and the former farmers were ordered to stand trial. On 8 May twenty-eight of the thirty-two accused were found guilty of conspiring against the people of France. Later that day they were taken in tumbrils from the Conciergerie to the Place de la Revolution. Lavoisier, the fourth to be guillotined, mounted the bloody scaffold after the rumbling blade had severed his father-in-law's head from his body. The victims' remains were unceremoniously buried in nameless graves in the cemetery of the Parc Monceaux.

Notes

ABBREVIATIONS

Corres. [fasc.:pp.] – Lavoisier, *Correspondance.*
D/K [item no.:pp.] – Duveen and Klickstein, *Bibliography.*
DSB [vol.:pp.] – *Dictionary of Scientific Biography.*
Grimaux – Grimaux, *Lavoisier.*
OL [vol.:pp.] – Lavoisier, *Oeuvres.*

INTRODUCTION

1 Baker, *French Revolution*, ch. 5.
2 As quoted in Heilbron, "Introductory essay," pp. 10–11. See also Bensaude-Vincent, "The balance: from chemistry to politics."
3 I have taken the phrase "rhetoric of numbers" from Lundgren, "Numbers in chemistry," pp. 259.

CHAPTER 1 *The Barristers of Paris*

1 The most comprehensive source for biographical information on Lavoisier is Grimaux; see ch. 1; see also McKie, *Lavoisier*, ch. 1.
2 Grimaux, pp. 383–4.
3 OL, III: 14, 84.
4 For geneological tables of the various branches of the Lavoisier family, see

Grimaux, pp. 325–30.

5   Bien, "The *Secrétaires du Roi*," pp. 154–5.

6   *Corres.*, 1: 2.

7   Grimaux. p. 9; Meldrum, 'Lavoisier's early work in science," p. 339.

8   OL, 11: 18; Gough, "Lavoisier's memoirs on the nature of water."

9   As quoted in Fitzsimmons, *Parisian Order of Barristers*, p. 1.

10 As quoted in Behrens, *The Ancien Régime*, p. 48.

11 The following description of the world of the barristers is based on Fitzsimmons, *Parisian Order of Barristers*, ch. 1.

12 See Shennan, *Parlement of Paris*; Stone, *The French Parlements*.

13 As quoted in Shennan, *Parlement of Paris*, pp. 309–10.

14 As quoted in ibid., p. 318.

15 Fitzsimmons, *Parisian Order of Barristers*, pp. 38–31 and n. 123. Meldrum, in "Lavoisier's early work in science," p. 352, notes that Lavoisier almost never cited the *Encyclopédie*, although he must have known it well.

16 As quoted in Fitzsimmons, *Parisian Order of Barristers*, p. 29.

Chapter 2   *The Republic of Science*

1   Torlais, *L'Abbé Nollet*, pp. 187, 192–4; Torlais, "La physique expérimentale," p. 627.

2   Heilbron, "Nollet," p. 147, n. 3.

3   Heilbron, *Electricity*, p. 287.

4   Nollet's inaugural lecture was published at the beginning of the first volume in later editions of his lectures on experimental physics; the passage quoted is from Nollet, *Leçons de physique expérimentale*, 5th edn (1759), vol. 1, p. lix.

5   Cf. Daston, "Republic of letters."

6   For a magisterial survey of the institutions that supported French science in the eighteenth century, see Gillispie, *Science and Polity*, esp. part 1.

7   The definitive study of the Paris Academy of Sciences in the seventeenth and eighteenth centuries is Hahn, *Anatomy of a Scientific Institution*.

8   Gillispie, "Carnot," p. 71.

9   Lavoisier, like a great many of his contemporaries, was also a Freemason. The Masons operated as a hierarchically organized voluntary society which, like the academies, provided its members with opportunities to engage in republican rituals of fraternal civility and self-governance in a setting protected from scrutiny by the government and the general public; see Roche, "Académies et politique," p. 337.

    For an extensive discussion of the consolidation of the French scientific community after the Terror, see Dhombres, *Naissance d'un pouvoir*.

10 Guerlac, *Essays and Papers*, pp. 327–33, esp. nn. 6 and 7.

11 Gingerich, "Lacaille."

12 For Lavoisier's fond recollection of La Planche as a teacher and his condemnation of the way students were introduced to the study of chemistry,
see Bensaude-Vincent, "The chemical revolution through contemporary
textbooks," p. 457. An account of La Planche's course is given in the
introduction to Beretta, "Lavoisier's first chemical paper."

13 Rappaport, "Guettard."

14 Guerlac, *Essays and Papers*, pp. 329–30.

15 Grimaux, p. 7. The Jansenists were severe Catholics who opposed the
Jesuits, supported the Gallican church, and were strong allies of the
parlements.

16 For further information on Lavoisier's geology and the publication of the
mineralogical atlas, see OL, V:1–238; D/K, no. 218:236–44; Duveen, *Supplement to a Bibliography*, pp. 129–32; Rappaport's papers, "Lavoisier's geologic activities," "Lavoisier and Monnet," "The geological atlas,"
"Lavoisier's theory of the earth."

17 Roche, "Académies et politique."

18 OL, VI:124; Grimaux, p. 150, n. 7, identifies the competition Lavoisier had
in mind when drafting this *éloge*.

19 This comparison is developed in Gillispie, *Science and Polity*, pp. 83–4.

20 Rappaport, "The liberties of the Paris Academy of Sciences," p. 252.

21 OL, IV:618.

22 Rappaport, "The liberties of the Paris Academy of Sciences," p. 231.

23 OL, III:14.

24 OL, III:1. On the change from collective and anonymous publication,
which had been the normal format in the Academy in the seventeenth
century, to individual publication in the eighteenth century, see Hahn,
Anatomy of a Scientific Institution, pp. 24–30.

25 OL, III:1–2.

26 OL, III:84.

27 Lavoisier's essay was not in fact published in his lifetime; see D/K, no.
343:385–6.

28 OL, III:111, n. 2. In taking up the study of gypsum, Lavoisier was addressing a set of analytic problems that had long been of central concern to
European chemists; see Holmes, *Investigative Enterprise*, lecture 2.

29 Hahn, *Anatomy of a Scientific Institution*, pp. 65, 111; Heilbron, *Electricity*,
p. 129; cf. Crow, *Painters and Public Life*, p. 33.

30 OL, III:106–44; D/K, no. 1:14–15, no. 348:387.

31 OL, III:106.

32 *Corres.*, 1:7–12.

33 This information comes from the volumes of MSS in the archives of the
Academy of Sciences that contain the transcribed minutes of the meetings
of the Academy (*procès-verbaux*). The chronologies compiled by Daumas
(*Lavoisier*, ch. 2) and Guerlac and Perrin ("Chronology") include numerous
excerpts from and references to these records. See ibid., p. 3.

NOTES TO PP. 40-50

34 *Corres.*, 1:80,92; Meldrum, "Lavoisier's early work in science", p. 344.
35 *Corres.*, 1:41.
36 *Corres.*, 1:49.
37 Evidently while in Strasbourg during the latter months of 1767 Lavoisier began identifying himself as a member of the Academy; see *Corres.*, 1:104; Guerlac, *Lavoisier*, p. 62.
38 It was not unusual for non-members to present papers to the Academy.
39 OL, III:427–50.
40 OL, III:145–205; D/K, no. 348:387; Guerlac and Perrin, "Chronology," p. 5.
41 OL, III:145.
42 Smeaton, "Venel," p. 602.
43 As quoted in Meldrum, "Lavoisier's early work in science," p. 348; on Lavoisier and Déparcieux, see ibid., pp. 345–8. See also OL, III:255; *Corres.*, 2:257–64.
44 OL, III:145.
45 Ibid., p. 146.
46 Guerlac and Perrin, "Chronology," p. 5; it was not in fact published, D/K, no. 348:387.
47 *Corres.*, 1:105–13.
48 *Corres.*, 1:114–17; Hahn, *Anatomy of a Scientific Institution*, p. 81; Rappaport, "The liberties of the Paris Academy of Sciences," p. 231.
49 *Corres.*, 1:117–18.

Chapter 3 *Experimental Physics*

1 On Lavoisier's pedagogical views see Daumas, *Lavoisier*, pp. 91–112; Bensaude-Vincent, "The chemical revolution through contemporary textbooks"; Beretta, *Lavoisier's First Chemical Paper.*
2 Translated from text published in Bensaude-Vincent, "The chemical revolution through contemporary textbooks," pp. 456–8.
3 Methodological diversity in eighteenth-century French chemistry is discussed in Donovan, "Newton and Lavoisier."
4 Berreta has dated, edited, and introduced the 19-page MS no. 380 in the Lavoisier papers deposited in the archives of the Academy of Sciences; see Beretta, *Lavoisier's First Chemical Paper.* I am grateful to Dr Beretta for sending me a draft of this publication.
5 For an account of the central importance of instrumentalism in experimental physics, the improvement of instruments in the eighteenth century, and 1760 as an important "point of inflection" in the rise of experimental physics, see Heilbron, "Introductory essay," pp. 1–10; on the role of experimental physics in eighteenth-century chemistry see Lundgren, "Numbers in chemistry," pp. 254–64.
6 The following account is taken from Beretta, *Lavoisier's First Chemical Pa-*

per.

7 *Corres.*, 1:9–10.

8 See Cohen, *Franklin and Newton*; Brunet, *Physiciens Hollandais*.

9 As quoted in Hahn, *Anatomy of a Scientific Institution*, p. 33; on late eighteenth-century positivism in France see Baker, *Condorcet*, ch. 3.

10 OL, III:113.

11 OL, III:450.

12 OL, III:262; for the dating of this document see Perrin, "Document, text and myth," pp. 7–10.

13 Perrin, "Lavoisier's thoughts on calcination and combustion," p. 664.

14 Ibid., p. 665.

15 Donovan, "Lavoisier as chemist."

16 Translated from quotation in Guerlac, *Crucial Year*, p. xvi.

17 Meldrum, "Lavoisier's early work in science," p. 332; Smeaton, "L'Avant-Coureur," p. 231.

18 Cf. Perrin, "Triumph of the antiphlogistians," p. 63.

19 Reprinted in Berthelot, *Révolution chimique*, pp. 46–9; Guerlac, *Crucial Year*, pp. 228–30.

20 OL, I:437–728; D/K, nos 121–5:94–119.

21 Cf. OL, V:267–70.

22 Smeaton, "The publication of Lavoisier's 'Mémoires de chemie' "; D/K, nos 186–200:199–214.

23 Heilbron, *Electricity*, ch. 1–2.

24 Diderot, "L'Interprétation de la nature," pp. 717–18.

25 See Shapin and Schaffer, *Leviathan and the Air-Pump*; Home, "Experimental physics"; Hankins, *Science and the Enlightenment*, ch. 3.

26 Donovan, "Buffon, Lavoisier"; Roger, "Buffon"; Gillispie, *Science and Polity*, pp. 143–51.

27 Buffon, *Oeuvres Philosopiques*, pp. 5–6.

28 Hanks, *Buffon*, pp. 37, 90.

29 Heilbron, *Electricity*, p. 347.

30 Roger, "Buffon," p. 577.

31 Hanks, *Buffon*, pp. 90–1.

32 Roger, "Diderot et Buffon."

33 Hanks, *Buffon*, pp. 92, 95; Roger, "Chimie et biologie."

34 Heilbron, *Electricity*, p. 347.

35 Ibid., pp. 346–7.

36 Gough, "Réaumur."

37 Lavoisier's response to Buffon's science was complicated by his early appreciation of Buffon's geological writings; see Rapapport, "The geological atlas," p. 281.

38 See Heilbron, *Electricity*, pp. 346–62; on Nollet's electrical theory see ibid., ch. 11.

39 See Hankins, *Science and the Enlightenment*, pp. 150–1.

40 Heilbron, *Electricity*, p. 349.
41 Ibid., p. 362.
42 Thackray, *Atoms and Powers*, pp. 205–18.
43 Crosland, "Development of chemistry," p. 427.
44 Buffon, *Oeuvres Philosophiques*, pp. 35–41; on the historic importance of this statement see Metzger, *Newton, Stahl, Boerhaave*, pp. 57–65.
45 Buffon, *Oeuvres Philosophiques*, p. 39.
46 Ibid., italics in original.
47 Ibid.
48 Itard, "Clairaut," pp. 282–3; Thackray, *Atoms and Powers*, pp. 155–61.
49 Macquer, *Dictionnaire de chymie*, art. "Pesateur," vol. 2, pp. 191–3.
50 Guyton, *Élémens de chymie*, vol. 1, p. 51; vol. 2, p. 5.
51 For citation of Geoffroy's table and a brief discussion of his ideas, see Smeaton, "Geoffroy."
52 On instruments in eighteenth-century physics see Heilbron, *Electricity*, pp. 71–3, 78–83.
53 The following quotations are taken from Nollet, *Leçons de physique expérimentale*, 5th ed. (1759), vol. 1, pp. iii–lxvii.
54 Alembert, "Expérimental"; see also Alembert's *Preliminary Discourse*, pp. 24–5.
55 Planck, *Autobiography*, pp. 33–4, as cited in Diamond, "The polywater episode," p. 198.
56 Kant, *Gesammelte Schriften*, vol. 4, pp. 470–1; as quoted in Gregory, "Romantic Kantianism," pp. 109, 113.
57 Heilbron, *Electricity*, pp. 61, 63.
58 On quantification see ibid., pp. 73–97.
59 Alembert presents similar views in his *Preliminary Discourse*.
60 Lacaille, *Leçons élémentaires de Mathématique*, p. 1.
61 On Lavoisier's use of mathematics see Holmes, *Lavoisier*, pp. 280–3, 289, 341–5, 371–2, 394, 396; Holmes, *Investigative Enterprise*, pp. 105–6.
62 OL, I:4.
63 OL, I:5.
64 OL, I:6. In the early 1780s, while working with Laplace, Lavoisier was somewhat more optimistic about the possibility of developing a well-grounded theory of chemical affinities; see Guerlac, "Chemistry as a branch of physics," pp. 266–76.
65 Jefferson, *Papers*, vol. 13, p. 381.

CHAPTER 4 *The Chemistry of Salts*

1 Hahn, *Anatomy of a Scientific Institution*, pp. 5, 16, 24–30, 59–60, 73.
2 The text of Lavoisier's presentation to the Easter meeting of 1773 is published in Fric, "Contribution," pp. 155–62. For the background to

Lavoisier's discovery and a thorough examination of the relevant documents see Guerlac, *Crucial Year*; Perrin, "Lavoisier's thoughts on calcination and combusion"; and Perrin, "Document, text and myth."

3   Fric, "Contribution," pp. 157–8.

4   Rappaport, "G. F. Rouelle"; Rappaport, "Rouelle and Stahl"; Fichman, "French Stahlism"; Metzger, *Newton, Stahl, Boerhaave.*

5   Venel, "Chymie."

6   Donovan, "Origins of modern chemistry"; Hannaway, *The Chemists and the Word*, pp. 155–6.

7   Rouelle's chemical lectures were never published, but many manuscript copies were and are available. I have relied primarily on the "Duveen" MS in the History of Science Collections of the Cornell University Library; Rouelle, "Traité de chymie," p. 7.

8   Ibid., p. 12.

9   Cf. Joseph Black's comments, as quoted in Donovan, *Philosophical Chemistry*, p. 144.

10   Rouelle developed the popular fixed-free theory of chemical elements at length in his lectures. For published discussions of this theory, see Venel, "Chymie"; Macquer, *Dictionnaire de chymie*, articles "Air", "Eau", "Feu"; Guerlac, *Crucial Year*, pp. 32–4, 94–8; Gough, "Lavoisier's memoirs on the nature of water," p. 98. For the background to Rouelle's fixed-free theory of salts see Holmes, *Investigative Enterprise*, ch. 2.

11   OL, III:123. On Nollet's memoir see Siegfried, "Lavoisier's view of the gaseous state," pp. 63–4.

12   As quoted in ibid.

13   Translation of quotation in Heilbron, *Electricity*, p. 77, n. 20.

14   Turgot, "Expansibilité"; on this topic see Gough, "Lavoisier's theory of the gaseous state," pp. 25–39.

15   On Rouelle's memoirs on salts see Rappaport, "G. F. Rouelle," p. 82; Partington, *History of Chemistry*, p. 74. See also Gough, "Lavoisier's theory of the gaseous state," p. 31, n. 60.

16   See Donovan, *Philosophical Chemistry*, p. 186.

17   OL, III:106–27.

18   Ibid., p. 114.

19   Ibid., pp. 106–7.

20   Meldrum, "Lavoisier's early work in science," pp. 341–2.

21   OL, III:128–44, p. 143.

22   Ibid., p. 128.

23   Ibid., p. 144.

24   Ibid., p. 123.

25   OL, III:61–70.

26   Ibid., p. 66.

27   Ibid., p. 67; cf. also p. 77.

28   See Siegfried, "Lavoisier's view of the gaseous state," p. 61.

29 Published, in French, in ibid., p. 62.

30 *Corres.*, 1:66.

31 Ibid., p. 72.

32 Ibid., p. 67.

33 Meldrum, "Lavoisier's early work in science," pp. 348–63; Gough, "Lavoisier's memoirs on the nature of water." For a brief history of the hydrometer and an account of Lavoisier's early interest in it see Beretta, *Lavoisier's First Chemical Paper*.

34 Lavoisier described his hydrometers in Part III of his *Elementary Treatise on Chemistry*; see Lavoisier, *Elements of Chemistry*, pp. 300–2 and plate 7, fig. 6.

35 OL, III:427–50.

36 Ibid., pp. 427, 428.

37 Ibid., pp. 448–9.

38 Ibid., pp. 449–50.

39 Ibid., pp. 145–205.

40 Ibid., pp. 146.

41 Ibid., pp. 148–9.

42 As translated in Meldrum, "Lavoisier's early work in science," p. 398.

43 OL, III:150.

44 OL, II:1–29, pp. 1–2; see also Gough, "Lavoisier's memoirs on the nature of water."

45 OL, II:6–7.

46 Ibid., p. 7.

47 See Gough, "Lavoisier's memoirs on the nature of water," pp. 92–7.

48 OL, II:25.

49 Daumas, *Lavoisier*, p. 27; Guyton de Morveau, *Digressions académiques*, pp. 1–267.

50 See Guerlac, *Crucial Year*, ch. 4, esp. pp. 136–45.

51 See ibid., pp. 79–90.

52 D/K, no. 9:21–2.

53 Guerlac, *Crucial Year*, pp. 55–9.

54 Smeaton, "Venel"; Perrin, "Document, text and myth," p. 12, n. 25.

55 Heilbron, *Electricity*, p. 434.

56 On Magellan and Trudaine see Guerlac, *Crucial Year*, pp. 36–55.

57 *Corres.*, 1:356–69.

58 Perrin, "Document, text and myth," p. 19; Holmes, "Lavoisier's conceptual passage."

59 Guerlac, *Crucial Year*, pp. 156–9.

60 OL, III:261–6; Guerlac, *Crucial Year*, pp. 88–95, 208–14; Perrin, "Document, text and myth," pp. 7–10.

61 Guerlac, *Crucial Year*, p. 209.

62 Ibid., p. 214.

63 Ibid., p. 94; OL, V:241–2.

64 Guerlac, *Crucial Year*, pp. 94–101, 215–23; Perrin, "Document, text and

myth," pp. 10–14.

65 Guerlac, *Crucial Year*, pp. 173–90, 223–7; Perrin, "Document, text and myth," pp. 14–16.

66 This account relies on Perrin, "Document, text and myth," pp. 18–25; and Perrin, "Lavoisier's thoughts on calcination and combustion," pp. 662–5.

67 Partington, *History of Chemistry*, p. 385, has pointed out that Lavoisier's account of this experiment is incomplete and suspect; William Smeaton, in a private communication, has suggested that in fact Lavoisier may have been describing a thought experiment.

68 *Corres.*, 2:389–90; Guerlac, *Crucial Year*, pp. 227–8.

69 Perrin, "Lavoisier's thoughts on calcination and combustion," p. 664.

70 Perrin, "Document, text and myth," pp. 17–18.

71 Perrin, "Lavoisier's thoughts on calcination and combustion," pp. 664–5. On Meyer's theory of fixed air see Guerlac, *Crucial Year*, p. 16; Gough, "Lavoisier's early career in science," pp. 54–5.

72 See above, ch. 3, n. 19.

73 OL, V:243–7; on date of presentation see Guerlac and Perrin, "Chronology," p. 13.

74 Smeaton, "L'Avant-Coureur."

75 OL, V:246–7.

76 See Guerlac and Perrin, "Chronology," p. 13.

77 See Gough, "Lavoisier's theory of the gaseous state," pp. 18–24; on Cullen experiments, see Donovan, *Philosophical Chemistry*, pp. 155–62.

78 Gough, "Lavoisier's theory of the gaseous state," p. 24.

79 OL, I:551–5.

80 OL, I:440–1.

CHAPTER 5 *The Company of Tax Farmers*

1 On the marriage and Lavoisier's wife's family, see Grimaux, pp. 35–60, 330–6, 381–4; see also McKie, *Lavoisier*, pp. 65–8.

2 Bosher, *French Finances*, p. 101.

3 In 1789 such a position cost 120,000 livres; Bien, "The *Secrétaires du Roi*," p. 154.

4 See Scheler, *Lavoisier*, pp. 40–3.

5 Perrin, "The Lavoisier–Bucquet collaboration," pp. 9–10.

6 See Schama, *Citizens*, ch. 2.

7 The following account is drawn primarily from Matthews, *Royal General Farms*; Durand, *Fermiers généraux*; Bosher, *French Finances*; Grimaux, pp. 62–82; OL, VI:125–85.

8 As quoted in Bien, "The *Secrétaires du Roi*," p. 156.

9 The following account is drawn largely from Grimaux, pp.32–4, 64–82; *Corres.*, 1:122–249, 2:250–320.

10 *Corres.*, 1:142, 167.
11 Ibid., 1:139–40, 152.
12 Ibid., 1:207.
13 Ibid., 1:193–4, 204.
14 Ibid., 1:231.
15 Ibid., 1:146.
16 Ibid., 1:204.
17 Ibid., 1:235–41; cf. OL, VI:134–5.
18 Gillispie, *Science and Polity*, pp. 64–5; see also Bensaude-Vincent, "The balance".
19 Durand, *Fermiers généraux*, p. 124.
20 *Corres.*, 2:298–9.
21 Ibid., 2:292; Grimaux, pp. 78–9.
22 OL, VI:109–24.

CHAPTER 6 *A New Theory of Combustion*

1 OL, II:283–333; D/K, no. 55:54–6. A fascimile reproduction of the original text and an English translation are available in Lavoisier and Laplace, *Memoir on Heat*. See also discussions of this memoir in Guerlac, "Chemistry as a branch of physics" and Roberts, "A word and the world."
2 Daumas, *Lavoisier*, p. 49. An expanded version of this report was read at the fall public meeting in 1783, see OL, II:334–59; D/K, no. 64:61–2. See also Guerlac, "Chemistry as a branch of physics", pp. 261–6.
3 Gillispie, *Invention of Aviation*, ch. 1–3; see also Gillispie, *Science and Polity*, pp. 535–44.
4 OL, II:623–55; D/K, no. 76:68–9. On the dating of this memoir see Daumas, *Lavoisier*, p. 58; Guerlac, "Chemistry as a branch of physics," p. 258. For a detailed analysis of this memoir see below, ch. 7.
5 OL, II:225–33; D/K, no. 45:48–9.
6 For detailed descriptions of Lavoisier's research from 1773 on see Berthelot, *Révolution chimique*, and Holmes, *Lavoisier*.
7 See Holmes, *Lavoisier*, pp. 27–40.
8 Priestley originally suspected that the unusual properties of the air he obtained from the red precipitate of mercury were a product of the nitric acid used in its formation; see Gibbs, *Priestley*, pp. 119–20.
9 See Guerlac, *Lavoisier*, ch. 6; Holmes, *Lavoisier*, ch. 2; Conant, "Phlogiston theory"; Perrin, "Prelude to Lavoisier's theory of calcination."
10 *Opuscules*, ch. 15, in OL, II:512.
11 As quoted in Conant, "Phlogiston theory," p. 83, from the memoir that Lavoisier read to the Easter public meeting of the Academy in 1775.
12 D/K, no. 29:33–8; no. 30:38–9; OL, II:122–8. Conant compares the two versions extensively in "Phlogiston theory," pp. 73–88.

13 As quoted, with minor changes, in ibid., p. 79.

14 As quoted in Holmes, *Lavoisier*, p. 18.

15 As quoted in Conant, "Phlogiston theory," p. 84.

16 Holmes, *Lavoisier*, p. 43; Perrin, "Early opposition to the phlogiston theory."

17 OL, II:512.

18 On Priestley see Golinski, *Science as Public Culture*, ch. 3 and 4; for an interesting contrast between French and English views of experiment see Dear, "Miracles."

19 As quoted from Priestley, *Experiments and Observations*, vol. 2, pp. 320–3, in Conant, "Phlogiston theory," p. 90.

20 As quoted in ibid., p. 90. This volume was published in December 1775 and advance sheets arrived in Paris in that month as well.

21 See Holmes, *Lavoisier*, pp. 16–17; Conant. "Phlogiston theory," pp. 74–6.

22 For a concise descrption of Priestley's chemical research program see Schofield, "Priestley," pp. 143–6, and Gibbs, *Priestley*.

23 Holmes, *Lavoisier*, p. 17.

24 Ibid.

25 On the eudiometer and its connection to the discovery of the composition of water see Partington, *History of Chemistry*, vol. 3, pp. 321–8; see also Golinski, *Science as Public Culture*, pp. 93–128, and Schaffer, "Measuring virtue."

26 OL, II:525–8.

27 Ibid., p. 528.

28 See Lavoisier's discussion of the analysis of the atmosphere in ch. 2 and 3 of his *Elementary Treatise on Chemistry*.

29 Conant, "Phlogiston theory," pp. 101–2. Priestley's experiments and reasoning were actually more complicated than this summary indicates; for full details see ibid., pp. 90–104.

30 See Guerlac, "Chemistry as a branch of physics," pp. 205–16, esp. p. 214.

31 OL, II:129–38.

32 Ibid., p. 130.

33 Ibid., p. 134.

34 Ibid., p. 137.

35 Ibid.

36 OL, II:248–60; D/K, no. 49:50–1. This memoir was not read until November 1779 and was not published until 1781. See also Crosland, "Lavoisier's theory of acidity."

37 OL, II:137–8.

38 Ibid., p. 138.

39 For a detailed study of the evolution of this memoir see Holmes, *Lavoisier*, ch. 3.

40 OL, II:174–5.

41 Ibid., p. 175.

42 Ibid., pp. 176–7.
43 Ibid., pp. 180–1.
44 See Holmes, *Lavoisier*, ch. 4.
45 For the reactions of Guyton de Morveau and Macquer to Lavoisier's proposed replacement theory see Guerlac, "Chemistry as a branch of physics," pp. 216–23.
46 OL, II:225–6.
47 Ibid., pp. 227–8.
48 Ibid., p. 228.
49 Ibid.; see also pp. 231–2.
50 Ibid., p. 231.
51 Ibid., p. 232.
52 Ibid., p. 231.
53 Ibid., p. 233.
54 Ibid.
55 OL, V:267–70; cf. Daumas, *Lavoisier*, p. 98.
56 See Guerlac, "Chemistry as a branch of physics," pp. 205–16; R. J. Morris, "Lavoisier and the caloric theory."
57 The instrument is described in Lavoisier's *Elementary Treatise on Chemistry*, Part III, ch. 3, and plate VI.
58 See Guerlac, "Chemistry as a branch of physics"; Holmes, *Lavoisier*, ch. 7.
59 See Guerlac, "Chemistry as a branch of physics," pp. 261–6; Perrin, "Lavoisier, Monge and the synthesis of water"; Partington, *History of Chemistry*, vol. 3, pp. 436–56.
60 Berthelot, *Révolution chimique*, p. 293; Holmes, *Lavoisier*, pp. 201–2.
61 The instrument is described in Lavoisier's *Elementary Treatise on Chemistry*, Part I, ch. VIII; Part III, ch. VIII, sect. VII; plate IV, fig. 5; see also Partington, *History of Chemistry*, vol. 3. p. 451. The instrument is preserved in the collection of Lavoisier materials at the Conservatoire des Arts et Métiers in Paris and is pictured in the David portrait of the Lavoisiers and in Kurzweil, "Laboratory of the soul," p. 90.
62 Translated from the letter as quoted in Guerlac, "Chemistry as a branch of physics," p. 264. Lavoisier also commented on the acid formed in the combustion of hydrogen and oxygen in a paper on the formation of nitric acid; see OL, V:610–11.
63 See Blagden's letter, reprinted in Partington, *History of Chemistry*, vol. 3, p. 441.
64 Perrin, "Lavoisier, Monge and the synthesis of water."
65 Priestley, *Scientific Autobiography*, pp. 215–20.
66 See Perrin, "Lavoisier, Monge and the synthesis of water," pp. 425–6; for another likely avenue of communication see Priestley's letter of January 1783 to Duluc in Paris, in Priestley, *Scientific Autobiography*, pp. 220–1.
67 On Lavoisier's later investigations of what Holmes calls "the chemistry of life" see Holmes, *Lavoisier*, ch. 10–17; on his further investigation of deto-

nation see Mauskopf, "Gunpowder and the chemical revolution."

68 See Guerlac, "Chemistry as a branch of physics," p. 266; Guerlac, *Lavoisier*, pp. 100–2. Priestley had also mentioned producing inflammable air from water and a heated gun barrel; Priestley, *Scientific Autobiography*, p. 221.

69 Daumas and Duveen, "Synthesis of water"; Langins, "Hydrogen production for ballooning."

CHAPTER 7 *The Campaign for French Chemistry*

1   OL, V:354–64; D/K, nos 126–152:119–54. Lavoisier's memoir was first published in Guyton de Morveau et al., *Nomenclature Chimique*, pp. 1–25.

2   Crosland, *Language of Chemistry*, pp. 184–6.

3   The committee's report, dated 18 June 1787, was published in Guyton de Morveau et al., *Nomenclature chimique*, pp. 238–52.

4   For the history of chemical nomenclature see Crosland, *Language of Chemistry*; for a philosophical treatment see Dagognet, *Langages de la chimie*. On Guyton see Smeaton's papers, "The reform of chemical nomenclature" and "Guyton de Morveau."

5   J. P. Marat was no doubt referring to Bergman and Guyton rather than Lavoisier when he wrote, in 1780, that "a famous author clamors against the nomenclature of chemistry"; see Gillispie, *Science and Polity*, pp. 309–11.

6   OL, V:355.

7   Translated from the quotation in Bensaude-Vincent, "A propos de *Méthode de nomenclature chimique*," pp. 21.

8   Ibid., p. 22.

9   OL, V:355–6.

10 As quoted in Baker, *French Revolution*, p. 116.

11 OL, V:356.

12 Condillac, *La Logique*. See also Albury, "The Logic of Condillac" and "The order of ideas." Lavoisier had long been interested in the language of chemistry and made reference to Condillac's philosophy of language from the early 1780s on; see Daumas, *Lavoisier*, pp. 99–112.

13 OL, V:361.

14 Ibid., p. 362.

15 On the battle for and against the new chemistry see Perrin, "The triumph of the antiphlogistians," esp. p. 54, for mention of a complaint that the new nomenclature treated the new theories as axioms.

16 Ibid., pp. 61–2; Perrin, "The chemical revolution." When arguing positivistically Lavoisier was prepared to leave differences over hypotheses unresolved; see OL, II:285–8, 332–3; see also Guerlac, "Chemistry as a branch of physics," pp. 244–50.

17 Le Grand, "The 'conversion' of C. L. Berthollet"; Sadoun-Goupil, *Berthollet*.

18 See Crosland, *Society of Arcueil*, pp. 232–48.

19 Smeaton, *Fourcroy*, ch. 2 and 3.

20 Hahn, *Anatomy of a Scientific Institution*, pp. 97–101; Perrin, "The triumph of the antiphlogistians," pp. 57–8; Smeaton, *Fourcroy*, pp. 27–8.

21 Siegfried, "Chemical revolution"; Perrin, "The triumph of the anti-phlogistians," pp. 59–61. On Fourcroy's changing views towards theories proposed by Lavoisier see Smeaton, *Fourcroy*, ch. 7.

22 Smeaton, *Fourcroy*, p. 229.

23 As quoted in Perrin, "The triumph of the antiphlogistians," pp. 41–2.

24 Smeaton, "Guyton de Morveau and the phlogiston theory."

25 Translated from quotation in Daumas, *Lavoisier*, pp. 60–1.

26 Perrin, "The triumph of the antiphlogistians," p. 46, n. 11.

27 OL, II:623–55. For dating see Daumas, *Lavoisier*, p. 58; Guerlac, "Chemistry as a branch of physics," p. 258.

28 OL, II:623.

29 Ibid., pp. 623–4.

30 Ibid., p. 624.

31 Ibid., p. 638.

32 Ibid., p. 640.

33 Ibid., pp. 640–1.

34 See Donovan, *Philosophical Chemistry*, pp. 265–77.

35 OL, II:644.

36 Ibid., p. 652.

37 Ibid., p. 653.

38 Perrin, "The triumph of the antiphlogistians," p. 46.

39 OL, II:655.

40 For France and Switzerland see Perrin, "The triumph of the anti-phlogistians"; for the spread of the new nomenclature see Crosland, *Language of Chemistry*. For Germany see Hufbauer, *German Chemical Community*; for Holland see Snelders, "The new chemistry in the Netherlands"; for Belgium see Halleux, "La révolution Lavoisienne en Belgique"; for Sweden see Lundgren, "The new chemistry in Sweden"; for Spain see Gago, "The new chemistry in Spain"; for Italy see Beretta, "A. L. Lavoisier en Italie"; for Scotland see Donovan, "Scottish responses," and Perrin, "Joseph Black"; for America see Greene, *American Science in the Age of Jefferson*, ch. 7; for Russia see Leicester, "Lavoisier in Russia." The reception in England was varied and has not been treated comprehensively, but on Priestley see Schofield, "Priestley," and McEvoy, "Continuity and discontinuity."

41 Perrin, "The triumph of the antiphlogistians," pp. 47–50.

42 Court, "The *Annales de Chimie*".

43 Bensaude-Vincent's "A propos de *Méthode de nomenclature chimique*" precedes a facsimile reprint of the entire book.

44 Kirwan, *Essay on Phlogiston*.

45 Perrin, "The triumph of the antiphlogistians," pp. 50–1.
46 Eyles, "Sir James Hall," pp. 167, 169–70.
47 OL, I:1–407, D/K, nos 153–85:154–99.
48 Lavoisier, *Elements of Chemistry*, p. xiii.
49 Daumas, *Lavoisier*, ch. 4; Bensaude-Vincent, "The chemical revolution through contemporary textbooks"; Beretta, *Lavoisier's First Chemical Paper*.
50 Daumas, *Lavoisier*, p. 101; Morris, "Lavoisier and the caloric theory," p. 2, n. 5.
51 Lavoisier, *Elements of Chemistry*, p. xxxiv.
52 Ibid., p. 176.
53 Ibid., p. 177.
54 Ibid., pp. 175–6. See Perrin, "Lavoisier's table of the elements"; Siegfried, "Lavoisier's table of simple substances."
55 Lavoisier, *Elements of Chemistry*, p. 210.
56 Ibid., p. xxxv.
57 On Lavoisier's instruments see Daumas, *Lavoisier*, ch. 5 and 6.
58 OL, I:101; cf. Lavoisier, *Elements of Chemistry*, p. 130.
59 Daumas, *Lavoisier*, p. 110.
60 Bensaude-Vincent, "The chemical revolution through contemporary textbooks."
61 OL, V:298–330.
62 OL, II:104.
63 See Dhombres, *Naissance d'un pouvoir*.
64 Daumas, *Lavoisier*, p. 109.
65 See Duveen and Klickstein, "Benjamin Franklin"; Fric, "Une lettre inédite."
66 Cf. Gough, "Some early references to revolutions in chemistry."
67 See Baker, *French Revolution*, ch. 9.
68 Translated from Grimaux, p. 126. On Chaptal's textbook see Bensaude-Vincent, "The chemical revolution through contemporary textbooks," pp. 441, 447–8. Lavoisier also continued to reach beyond the circle of those who accepted his new theories. In July 1791 Joseph Priestley organized a celebration in Birmingham to mark the anniversary of the fall of the Bastille. Later that day a Church and King crowd destroyed his home, library, and laboratory. Priestley's French colleagues soon sent messages of sympathy and support. The letter Lavoisier wrote for the chemists of Paris contains passages that convey perfectly his vision of the republic of science, the primacy of experimental investigation, and the political significance of science.

As a Citizen, you [i.e. Priestley] belong to England, and it is for her to atone for your losses: as a Scholar and as a Philosopher you belong to the entire world; you belong, above all, to those who know how to appreciate you, and it is we, united in agreement, who vow to restore to you the instruments which you have employed so usefully in our instruction. We have therefore resolved to reestablish your Cabinet, to raise again the Temple which ignorance, barbarity, and superstition have dared profane. What more important service can we render to science than to place in your hands the instruments necessary for its cultivation?

Zealous defender of the liberty which our country aims to obtain, we have not been idle witnesses of the most astonishing of the revolutions of which the annals of the world have preserved the memory. We have followed its movements and progress with the spirit of observation which the study of science bestows and which we have imbibed from the works of our teacher.

As quoted in Priestley, *Scientific Autobiography*, p. 258. On the dating of this letter and Lavoisier's authorship see Beretta, "Chemists in the storm," n. 73; I am grateful to Dr Beretta for sending me a preprint of this publication.

69 On chemistry as the pardigmatic progressive science in the first two decades of the nineteenth century see Golinski, *Science as Public Culture*, p. 236. Brief discussions of the second scientific revolution can be found in Gillispie, *Science and Polity*, p. 74, and Cohen, *Revolution in Science*, pp. 91–2; see also Donovan, "Newton and Lavoisier."

70 For a study of someone who resisted this revolution see Burlingame, "Lamarck's chemistry."

71 Fox, "Laplacian physics."

72 Kremer, "Defending Lavoisier"; Bensaude-Vincent, "Mythologie révolutionnaire"; Bensaude-Vincent, "A founder myth in the history of science."

CHAPTER 8 *Gunpowder and Agriculture*

1 See Dakin, *Turgot*, and Fauré, *La Disgrâce de Turgot*; see also Schama, *Citizens*, pp. 79–87.

2 See descriptions in Schama, *Citizens*, pp. 51–4; Baker, *French Revolution*, pp. 109–12.

3 Fox-Genovese, *Origins of Physiocracy*.

4 Kaplan, *Bread, Politics, and Political Economy*.

5 In 1789, while serving as a member of the Assembly of the Nobility of Blois, Lavoisier donated 50,000 livres to the city's bakers so they could buy flour during a famine; Scheler, *Lavoisier*, p. 103. See also Greenbaum, "Humanitarianism of Antoine Laurent Lavoisier."

6 Grimaux, p. 84.

7 McKie, *Lavoisier*, p. 92.

8   OL, V:392, 605–13. For the early history of gunpowder see Partington, *Greek Fire and Gunpowder*; on eighteenth-century theories of detonation, and Lavoisier's investigations of this problem, see Mauskopf, "Gunpowder and the chemical revolution"; see also Mauskopf, "Chemistry and cannon."

9   OL, V:391–460; see also Multhauf, "The French crash program for saltpeter production" and Gillispie, *Science and Polity*, pp. 50–3.

10 On the administrative history of gunpowder production see OL, V:680–702; Payen, *L'Evolution d'un monopole*; Gillispie, *Science and Polity*, pp. 50–8; Scheler, "Lavoisier et la Règie des Poudres."

11 Grimaux, p. 85.

12 OL, V:461–72; Grimaux, p. 86; Gillispie, *Science and Polity*, p. 58, n. 163.

13 Ibid., p. 68.

14 D/K, nos 302–16:219–35.

15 OL, V:494.

16 Gillispie, *Science and Polity*, pp. 69–70.

17 Ibid., pp. 69, 72–3; OL, V:498–604.

18 Gillispie, *Science and Polity*, p. 68; Multhauf, "The French crash program for saltpeter production," p. 169.

19 Partington, *History of Chemistry*, vol. 3, pp. 510–11.

20 Lavoisier's account of this episode was published in the *Journal de Paris*; see OL, V:742–5; see also Gillispie, *Science and Polity*, pp. 70–2.

21 See also Storrs, "Lavoisier's technical reports"; Duveen and Klickstein, "Lavoisier's contributions to medicine and public health."

22 OL, V:700.

23 OL, V:680–92.

24 OL, V:687.

25 Ibid., p. 681. On the intense public debate over bureaucracy and central authority in the last years of Louis XV's reign see Baker, *French Revolution*, ch. 7, esp. pp. 160–3; on revolutionary complaints that royalism had restricted production of saltpeter, and on the establishment of free enterprise in this trade in 1793, see Multhauf, "The French crash program for saltpeter production," p. 173.

26 Gillispie, *Science and Polity*, pp. 360–88, esp. p. 365. See also Bourde, *Agronomie et agronomes*; Lenglen, *Lavoisier agronome*; Schelle and Grimaux, *Lavoisier*.

27 Grimaux, pp. 149–50.

28 Fox-Genovese, *Origins of Physiocracy*, p. 13; Gillispie, *Science and Polity*, ch. 1 and passim. On Dupont see Saricks, *Du Pont*. The *physiocrates* were *philosophes* who appealed to economic principles when constructing rational systems of taxation and proposing programs for administrative reform.

29 Smeaton, "The publication of Lavoisier's 'Mémoires de chimie'", pp. 26–7.

30 There is a picture of the house in Grimaux, p. 167.

31 OL, II:812–23; summarized in McKie, *Lavoisier*, pp. 203–17.
32 Smeaton, "Lavoisier's membership of the Société Royale d'Agriculture," pp. 270–2; cf. Rappaport, "The geological atlas".
33 Gillispie, *Science and Polity*, pp. 379–87; Smeaton, "Lavoisier's membership of the Société Royale d'Agriculture."
34 While serving as the administrator of the Academy of Sciences in 1785, Lavoisier carried through a reorganization and expansion of the scientific sections. As part of this reform, the study of agriculture was added as a new area of concern within the botany section; see Gillispie, *Science and Polity*, p. 371.
35 The minutes Lavoisier kept as secretary of the Committee have been published in Pigeonneau and de Foville, *L'Administration de l'agriculture*.
36 OL, VI:227–9.
37 Ibid., pp. 403–16; cf. Dujarric de la Rivière, *Lavoisier économiste*.
38 OL, VI:407.
39 Ibid., pp. 404–5.
40 Ibid., pp. 415–16.
41 Ibid., p. 405.
42 Ibid.
43 Ibid., p. 416.
44 Grimaux, p. 237.
45 Cf. ibid., p. 78.

CHAPTER 9 *Mesmerism and Public Opinion*

1 The following account draws primarily on OL, III:499–527; Gillispie, *Science and Polity*, pp. 258–89; Darnton, *Mesmerism*; Sutton, "Electrical medicine and mesmerism." For illustrations of mesmerist sessions, see Darnton, pp. 7, 9. See also D/K, nos 80–1:71–3; nos 223–232b:249–61; no. 369:392.
2 Police reports of the 1780s explicitly warned that mesmerism was seditious; see Darnton, *Mesmerism*, pp. 83–8.
3 Gillispie, *Science and Polity*, pp. 276–80.
4 OL, III:514.
5 While Bailly prepared the committee's final report, Lavoisier was very active in formulating the committee's strategy and directing its affairs; Grimaux, p. 132.
6 See the letter quoted in Grimaux, p. 134 for evidence that Breteuil, Lavoisier, and Bailly also employed ridicule to weaken popular belief in mesmerism.
7 OL, III:505–6; Gillispie, *Science and Polity*, p. 277; Darnton, *Mesmerism*, p. 52.
8 Translated in ibid., pp. 189–92; cf. Baker, *Condorcet*, pp. 76–80, and Hahn, *Anatomy of a Scientific Institution*, p. 158.
9 The naturalist Jussieu submitted a minority report that was considerably

more open-minded; Gillispie, *Science and Polity*, pp. 283–4.

10 The view that mesmerism was a danger to public health was widely shared: in the spring of 1784 the journalist La Harpe wrote that it was "an epidemic that has overcome all of France"; Darnton, *Mesmerism*, p. 40.

11 OL, III:508.

12 Ibid., p. 509.

13 Ibid.

14 Ibid., p. 516.

15 Ibid., p. 517.

16 Some of these experiments were performed at Lavoisier's house; Darnton, *Mesmerism*, p. 64.

17 OL, III:521–2.

18 Darnton, *Mesmerism*, pp. 62–72. See also Crow, *Painters and Public Life*, pp. 224–6, on the subsequent Kornmann affair.

19 As quoted in Darnton, *Mesmerism*, p. 66.

20 As quoted in Gillispie, *Science and Polity*, p. 289.

21 The following account is largely based on ibid., pp. 290–330; see also Gottschalk, *Marat*.

22 Gillispie, *Science and Polity*, pp. 299–300; Popkin, "Marat and the eighteenth-century science of violence."

23 A picture obtained with this device in reproduced in Gillispie, *Science and Polity*, p. 311.

24 As quoted in ibid., p. 307.

25 Ibid., p. 316.

26 OL, IV:360.

27 The following account of Brissot is drawn largely from Darnton, *Mesmerism*, and Darnton, *Literary Underground*, pp. 41–70. See also Ellery, *Brissot de Warville*, and Hampson, *Will and Circumstance*.

28 See Darnton, *Literary Underground*, pp. 1–40.

29 As quoted in Baker, *Condorcet*, p. 76.

30 For a discussion of their attacks on the Academy see Hahn, *Anatomy of a Scientific Institution*, pp. 150–8.

31 Darnton, *Mesmerism*, p. 92.

32 Gillispie, *Science and Polity*, pp. 314–15.

33 See Baker, *Condorcet*, ch. 8; Darnton, *Literary Underground*, pp. 82–7.

34 As quoted in ibid., p. 75.

35 See Darnton, *Mesmerism*, p. 79; see also Guerlac, "Commentary," p. 319.

36 Gillispie, *Science and Polity*, p. 329.

37 Translated from quotation in Grimaux, p. 207. Marat also attacked Lavoisier in his journal *L'Ami du peuple* on 27 Jan. 1791; see ibid., pp. 206–7.

38 See Schama, *Citizens*, ch. 5.

39 On Bergasse see Darnton, *Mesmerism*, pp. 101–3.

40 This account of the wall is based primarily on Matthews, *Royal General*

Farms, pp. 168–73; see also Grimaux, pp. 80–2; McKie, *Lavoisier*, pp. 129–30. In 1767 Lavoisier recalled that it was Turgot who, during his brief ministry, first proposed building a new wall and asked Lavoisier to prepare a report on the subject; see Beretta, "Chemists in the storm," p. 43

41 As quoted from Mercier's *Tableau de Paris* in Matthews, *Royal General Farms*, p. 277.

42 Kennedy, *Cultural History of the French Revolution*, pp. 5–13.

43 Critics of the wall claimed that over 30 million livres were spent on its construction; Grimaux, p. 80; Matthews reports that slightly under 6½ million liveres were appropriated for its completion; *Royal General Farms*, p. 173.

44 Grimaux, pp. 80–1.

Chapter 10 *Representation, Legislation, and Finance*

1 Grimaux, pp. 167–86. Qualifications for representation were, by later standards, rather permissive at this time and it was not unusual for Lavoisier to serve on one occasion as a representative of the third estate and, in a later assembly, as a representative of the nobility.

2 McKie, *Lavoisier*, p. 212; Smeaton, "Monsieur and Madame Lavoisier in 1789," p. 4, n. 8.

3 Baker, *French Revolution*, p. 240.

4 See ibid., pp. 238–43; Baker, *Condorcet*, pp. 202–14, 244–60; Cavanaugh, "Turgot."

5 As quoted in Baker, *Condorcet*, p. 203.

6 Ibid., p. 207.

7 See Doyle, *French Revolution*, p. 3, for a map that distinguishes between the *pays d'états* and the *pays d'élections*. In the *pays d'états* (Brittany, Bayonne, Languedoc, Provence, Dauphiné, Burgundy, Franch-Comté, Alsace, Lorraine, Artois, and Flanders), representation was already assured through existing provincial estates.

8 OL, VI:238–312.

9 Doyle, *French Revolution*, pp. 84–93; Grimaux, pp. 191–2.

10 OL, VI:313–34; cf. Grimaux, pp. 191–5.

11 Baker, *Condorcet*, p. 252–5.

12 From a report by Mirabeau of a comment to Dupont, as quoted in Cavanaugh, "Turgot," p. 42; cf. Baker, *Condorcet*, p. 305.

13 As quoted in Cavanaugh, "Turgot," p. 58, n. 98; cf. Baker, *Condorcet*, p. 211.

14 OL, VI:313; cf. Cavanaugh, "Turgot," p. 56, on Turgot's fundamental republicanism.

15 OL, VI:314.

16 Ibid., p. 317.

17 Ibid., p. 319.

18 Ibid., p. 320.
19 Ibid., pp. 320–1.
20 Ibid., p. 321.
21 Ibid., pp. 322–3.
22 Ibid., p. 332.
23 Ibid., p. 334.
24 Doyle, *French Revolution*, p. 89.
25 As quoted in Baker, *Condorcet*, p. 207.
26 OL, VI:323.
27 Doyle, *French Revolution*, p. 93–7.
28 Ibid., p. 97.
29 Grimaux, pp, 195–8; OL, VI:335–63; D/K, nos 246–7:281–4.
30 Scheler, *Lavoisier*, p. 103.
31 OL, VI:335.
32 Ibid., pp. 351–3.
33 Ibid., p. 344.
34 Ibid., p. 358.
35 D/K, no. 248:284–6; Scheler, *Lavoisier*, p. 103.
36 Bosher, *French Finances*, ch. 14 and 15.
37 D/K, nos 249–55:286–90; Bigo, *La Caisse d'escompte*.
38 As quoted in Doyle, *French Revolution*, p. 105.
39 As quoted in ibid., p. 108.
40 See ibid., ch. 4.
41 Ibid., ch. 5.
42 Ibid., p. 116.
43 Grimaux, pp. 200–1.
44 Doyle, *French Revolution*, p. 131.
45 As translated in Guerlac, *Lavoisier*, pp. 127–8.
46 As quoted in McKie, *Lavoisier*, p. 222.
47 Bosher, *French Finances*, pp. 268–9. On the Society of 1789 see Baker, *Condorcet*, pp. 272–85. For Lavoisier's memoir see OL, VI:364–8; summarized in McKie, *Lavoisier*, pp. 224–7.
48 OL, VI:370.
49 Lavoisier, good bourgeois and patriot that he was, eagerly purchased nationalized lands as they came on the market. Scheler, *Lavoisier*, pp. 145–6, provides evidence that he spent over $1\frac{1}{4}$ million livres while acquiring extensive agricultural properties.
50 Schama, *Citizens*, p. 708.
51 Grimaux, pp. 209–11; Scheler, *Lavoisier*, pp. 114–17. On the development of the Treasury, see Bosher, *French Finances*, ch. 13.
52 Doyle, *French Revolution*, p. 158.
53 OL, VI:464–515.
54 Ibid., p. 464.
55 Ibid., p. 465.

56 Grimaux, p. 212.
57 Grimaux, pp. 215–16.
58 As quoted in Doyle, *French Revolution*, p. 193.
59 Bosher, *French Finances*, pp. 274–5.

CHAPTER 11 *The Republic of Virtue*

1   OL, V:721–2; D/K, no. 320:328–30. See also Scheler, *Lavoisier et la Révolution française*, vol. 2, pp. 120–9, 132; Grimaux, pp. 93–6; Smeaton, "Lavoisier's membership of the Assembly of Representatives," pp. 236–8.
2   Translated from a letter quoted in Scheler, *Lavoisier et la Révolution française*, vol. 2, p. 126.
3   See Chartier, *Cultural Origins*, chs. 1 and 2.
4   On David and the salons see Crow, *Painters and Public Life*, ch. 7; Dowd, *Pageant-Master of the Republic*, ch. 1. Lavoisier attended the salon in 1785 and made some notes on David's *Horatii*; see Grimaux, pp. 378–80.
5   Crow, *Painters and Public Life*, pp. 245–54.
6   Dowd, *Pageant-Master of the Republic*, p. 19.
7   As quoted in Gillispie, *Science and Polity*, p. 64.
8   Daumas, *Lavoisier*, ch. 6; Holmes, *Lavoisier*, pp. 403–9; Levere, "Balance and gasometre." See also Daumas, "Appareils d'experimentation," and Daumas, *Les Instruments scientifiques*.
9   Holmes, *Lavoisier*, ch. 14; Holmes, *Investigative Enterprise*, ch. 6.
10 Holmes, *Lavoisier*, ch. 16.
11 Ibid., pp. 443–4 and n. 17.
12 OL, II:68.
13 As quoted in Holmes, *Lavoisier*, p. 459.
14 Ibid.
15 See Hahn, *Anatomy of a Scientific Institution*, ch. 6 and pp. 324–5.
16 Lavoisier, although not on the committee, was clearly much involved; cf. OL, IV:597–614. As Hahn has pointed out (*Anatomy of a Scientific Institution*, p. 170, n. 31), this document was misplaced by the editors.
17 Ibid., pp. 190–4. See also the list of members given in Grimaux, p. 246, n. 1; and Gillispie, "Jacobin philosophy of science," pp. 270–9.
18 Gillispie, ibid., p. 278, says he was probably one of its organizers. From 2 October 1793 until his imprisonment on 28 November 1793, Lavoisier was president of the Advisory Office; Grimaux, p. 254; see also OL, IV:624–715.
19 OL, IV:649–68; VI:516–58; see also Baker and Smeaton, "Education proposals." On the debates over this plan see Hahn, *Anatomy of a Scientific Institution*, ch. 7.
20 Baker and Smeaton, "Education proposals," p. 46.
21 See Dhombres, *Science et savants en France*.
22 As quoted in Hahn, *Anatomy of a Scientific Institution*, p. 182.

23 Smeaton, "The Lycée and the Lycée des Arts."
24 Hahn, *Anatomy of a Scientific Institution*, pp. 232–3.
25 OL, VI:559–69.
26 As quoted in Baker and Smeaton, "Education proposals," p. 34.
27 As quoted in Hahn, *Anatomy of a Scientific Institution*, p. 184; the work quoted was published in 1791.
28 Ibid., p. 185.
29 As quoted in ibid., p. 183. For an admirably provocative argument that the Jacobins did indeed articulate a distinctive epistemology, see Gillispie, "Jacobin philosophy of science."
30 See Hahn, *Anatomy of a Scientific Institution*, ch. 8.
31 OL, VI:660–712; Grimaux, pp. 218–22; Hahn, *Anatomy of a Scientific Institution*, pp. 162–4. See also Fayet, *Révolution Française*, chs. 15 and 16; Heilbron, "The measure of enlightenment."
32 Hahn, *Anatomy of a Scientific Institution*, p. 227.
33 Grimaux, p. 224.
34 The following account and quotations are from Hahn, *Anatomy of a Scientific Institution*, pp. 212–13.
35 Fayet, *Révolution Française*, pp. 105–8.
36 OL, IV:615–24; Hahn, *Anatomy of a Scientific Institution*, ch. 8.
37 OL, IV:616.
38 Ibid., p. 618.
39 Ibid., pp. 621–2.
40 Hahn, *Anatomy of a Scientific Institution*, pp. 237–40.
41 Ibid., pp. 245–8; Fayet, *Révolution Française*, pp. 152–61.
42 Grimaux, pp. 208–14.
43 On Lavoisier's final publication projects see OL, II:99–104; D/K, nos 186–200:199–214; Bensaude-Vincent, "The chemical revolution through contemporary textbooks," pp. 448–51, 456–60; Smeaton, "The publication of Lavoisier's 'Mémoires de chimie'." The social and psychological effects of the Terror on the French scientific community are examined in Outram, "Ordeal of vocation."
44 See Nora, "Republic."
45 Furet, "Terror," p. 137.
46 Morris, *Diary of the French Revolution*, vol. 1, p. 243. Cf. Madame Lavoisier's description of their life during the early months of the Revolution, in a letter dated 1 October 1789 to the Italian chemist Landrioni; Beretta, "Chemists in the Storm," appendix.
47 Grimaux, p. 200.
48 Grimaux, pp. 276–9, 288–9, 298–9; OL,IV:713–15; Duveen, "Lavoisier and the French Revolution," part 3.
49 Kersaint, *Fourcroy*, pp. 72–4; Smeaton, *Fourcroy*, pp. 58–9.
50 Full details can be found in Grimaux, ch. 8; McKie, *Lavoisier*, ch. 25.
51 Pouchet, in *Les sciences pendant la terreur*, p. 12, reports that there were no

scientists among the émigrés who fled France during the Revolution. It is interesting to note that, in a review of a biography of Werner Heisenberg, Walter J. Moore, the author of a biography of Erwin Schrödinger, wrote that,

> Heisenberg's loyalty to the Nazi regime was not anomalous. There is no evidence that any German professor of physics or chemistry who did not have a Jewish family ever expressed public disapproval of Hitler or left Germany because of the atrocities (though it should be noted that the Austrian physicist Erwin Schrödinger quit his Berlin professorship in disgust as soon as Hitler took power). Nor is there anything peculiarly Germanic in Heisenberg's patriotic attitude of "my country, right or wrong." Scientists under all regimes tend to prefer expediency to morality.

(*New York Times Book Review*, 8 March 1992, p. 11.)

52 Grimaux, pp. 385–6.
53 Ibid., p. 276.
54 Ibid., pp. 297–8.
55 For accounts of her life see ibid., pp. 331–6; Duveen, "Madame Lavoisier."
56 G. Morris, *Diary of the French Revolution*, vol. 1, p. 109.
57 Saricks, *Du Pont*, p. 289, n. 73; Scheler, *Lavoisier*, pp. 147–8.
58 Grimaux, pp. 289–90.

Bibliography

Alembert, Jean d', "Expérimental," in *Encyclopédie, ou Dictionnaire raisoné des sciences, des arts et des métiers, par une société de gens de lettres*, ed. Denis Diderot and Jean d'Alembert, vol. 6, Paris, 1756, pp. 298–301.

—— *Preliminary Discourse to the Encyclopedia of Diderot*, trans. Richard N. Schwab, Bobbs-Merrill, Indianapolis, 1963.

Albury, W. R., "The Logic of Condillac and the Structure of French Chemical and Biological Theory, 1780–1801," Ph.D. dissertation, Johns Hopkins University, Baltimore, 1972.

—— "The order of ideas: Condillac's method of analysis as a political instrument in the French revolution," in *The Politics and Rhetoric of Scientific Method*, ed. John A. Schuster and Richard Yeo (Australasian Studies in History and Philosophy of Science, vol. 4), Reidel, Dordrecht, 1986, pp. 203–25.

Baker, Keith Michael, *Condorcet*, University of Chicago Press, Chicago, 1975.

—— *Inventing the French Revolution*, Cambridge University Press, Cambridge, 1990.

Baker, Keith Michael, and Smeaton, W. A., "The origins and authorship of the education proposals published in 1793 by the *Bureau de Consultation des Arts et Métiers* and generally ascribed to Lavoisier," *Annals of Science*, 21 (1965), pp. 33–46.

Behrens, C. B. A., *The Ancien Régime*, Harcourt, Brace and World, New York, 1967.

Bensaude-Vincent, Bernadette, "The balance: between chemistry and politics," *The Eighteenth Century: Theory and Interpretation*, 33 (1992), pp. 217–37.

—— "A founder myth in the history of science? – the Lavoisier case," in *Functions and Uses of Disciplinary Histories*, ed. Loren Graham, Wolf Lepenies and Peter Weingart (Sociology of the Sciences Yearbook, vol. 7), Reidel,

Dordrecht, 1983, pp. 53–78.

—— "Une mythologie révolutionnaire dans la chimie Française," *Annals of Science*, 40 (1983), pp. 189–96.

—— "A propos de *Méthode de nomenclature chimique*," in *Cahiers d'Histoire et de Philosophie des Sciences*, new series, no. 6, Centre National de la Recherche Scientifique, Paris, 1983, pp. 1–39.

—— "A view of the chemical revolution through contemporary textbooks: Lavoisier, Fourcroy and Chaptal," *British Journal for the History of Science*, 23 (1990), pp. 435–60.

Beretta, Marco, "A. L. Lavoisier en Italie (1774–1800)," in *Échanges d'influences scientifiques et techniques entre pays européens de 1780 à 1830*, Actes du 114e congrés national des sociétés savantes, Éditions du CTHS, Paris, 1990, pp. 125–44.

—— "Chemists in the storm: Lavoisier and Priestley and the French Revolution," *Nuncius*, 7 (1993), in press.

—— *A New Course in Chemistry: Lavoisier's First Chemical Paper*, forthcoming.

Berthelot, Marcellin, *La révolution chimique – Lavoisier*, Alcan, Paris, 1890.

Bien, David D., "The *Secrétaires du Roi*: absolutism, corps and privilege under the ancien régime," in *Vom Ancien Régime zur Französischen Revolution*, ed. Ernest Hinrichs, Eberhard Schmitt, and Rudolf Vierhaus, Vandenhoeck & Ruprecht, Göttingen, 1978, pp. 153–68.

Bigo, R., *La Caisse d'escompte (1776–1793) et les origins de la Banque de France*, Presses Universitaires de France, Paris, 1927.

Bosher, J. F., *French Finances 1770–1795 – From Business to Bureaucracy*, Cambridge University Press, Cambridge, 1970.

Bourde, André, *Agronomie et agronomes en France au XVIIIe siècle*, 3 vols, S.E.V.P.E.N., Paris, 1967.

Brunet, Pierre, *Physiciens Hollandais et la méthode expérimentale en France au XVIIIe siècle*, Blanchard, Paris, 1926.

Buffon, *Oeuvres philosophiques*, ed. Jean Piveteau, Presses Universitaires de France, Paris, 1954.

Burlingame, Leslie J., "Lamarck's chemistry: the chemical revolution rejected," in H. Woolf, ed., *The Analytic spirit*, pp. 64–81.

Cavanaugh, Gerald J., "Turgot: the rejection of enlightened despotism," *French Historical Studies*, 6 (1969), pp. 31–58.

Chartier, Roland, *The Cultural Origins of the French Revolution*, trans. Lydia G. Cochrane, Duke University Press, Durham, North Carolina, 1991.

Cohen, I. Bernard, *Franklin and Newton*, American Philosophical Society, Philadelphia, 1956.

—— *Revolution in Science*, Harvard University Press, Cambridge, 1985.

Conant, James Bryant, "The overthrow of the phlogiston theory: the chemical revolution of 1775–1789," in *Harvard Case Histories in Experimental Science*, ed. James Bryant Conant and Leonard K. Nash, 2 vols., Harvard University Press, Cambridge, 1957, vol. 1, pp. 65–115.

Condillac, Abbé de, *La Logique, ou les primiers développemens de l'art de penser*, Paris, 1780.

Court, Susan, "The *Annales de Chimie*, 1789–1815," *Ambix*, 19 (1972), pp. 113–28.

Crosland, Maurice P., "The development of chemistry in the eighteenth century," *Studies on Voltaire and the Eighteenth Century*, 24 (1963), pp. 369–441.

—— *Historical Studies in the Language of Chemistry*, Harvard University Press, Cambridge, 1962.

—— "Lavoisier's theory of acidity," *Isis*, 64 (1973), pp. 306–25.

—— *The Society of Arcueil: a view of French science at the time of Napoleon I*, Harvard University Press, Cambridge, 1967.

Crow, Thomas E., *Painters and Public Life in Eighteenth-Century Paris*, Yale University Press, New Haven, 1985.

Dagognet, F., *Tableaux et langages de la chimie*, Seuil, Paris, 1969.

Dakin, Douglas, *Turgot and the Ancien Régime in France*, Methuen, London, 1939.

Darnton, Robert, *The Literary Underground of the Old Regime*, Harvard University Press, Cambridge, 1982.

—— *Mesmerism and the End of the Enlightenment in France*, Harvard University Press, Cambridge, 1968.

Daston, Lorraine, "The ideal and reality of the republic of letters in the enlightenment," *Science in Context*, 4 (1991), pp. 367–86.

Daumas, Maurice, "Les appareils d'experimentation de Lavoisier," *Chymia*, 3 (1950), pp. 45–62.

—— *Les Instruments scientifique aux XVIIe et XVIIIe siècles*, Presses Universitaires de France, Paris, 1953.

—— *Lavoisier. Theoreticien et Experimentateur*, Presses Universitaires de France, Paris, 1955.

Daumas, Maurice, and Duveen, Denis, "Lavoisier's relatively unknown large-scale decomposition and synthesis of water, February 27 and 28, 1785," *Chymia*, 5 (1959), pp. 113–29.

Dear, Peter, "Miracles, experiments, and the ordinary course of nature," *Isis*, 81 (1991), pp. 663–83.

Dhombres, Nichole, et Dhombres, Jean, *Naissance d'un pouvoir: sciences et savants en France (1793–1824)*, Payot, Paris, 1989.

Diamond, Arthur M., Jr., "The polywater episode and the appraisal of theories," in *Scrutinizing Science*, ed. Arthur Donovan, Larry Laudan and Rachel Laudan, Kluwer, Dordrecht, 1988, pp. 181–98; reprint, Johns Hopkins University Press, 1992.

Dictionary of Scientific Biography, ed. Charles Coulston Gillispie, 16 vols., Scribner's, New York, 1970–80.

Diderot, Denis, "Pensées sur l'interprétation de la nature," in D. Diderot, *Oeuvres complètes*, édition chronologique, Club français du livre, Paris, vol. 2 (1969), pp. 708–74.

Donovan, Arthur, "Buffon, Lavoisier and the transformation of French chemistry," in *Buffon 88: Actes du colloque international*, ed. J. Gayon, Vrin, Paris, 1992,

pp. 387–95.
—— "Lavoisier and the origins of modern chemistry," in A. Donovan, ed., *Chemical Revolution*, pp. 214–31.
—— "Lavoisier as chemist *and* experimental physicist," *Isis*, 81 (1990), pp. 270–2.
—— "Newton and Lavoisier – from chemistry as a branch of natural philosophy to chemistry as a positive science," in *Action and Reaction*, ed. Paul Theerman and Adele F. Seeff, Newark, University of Delaware Press, 1993, pp. 255–76.
—— *Philosophical Chemistry in the Scottish Enlightenment*, Edinburgh University Press, Edinburgh, 1975.
—— "Scottish responses to the new chemistry of Lavoisier," in *Studies in Eighteenth-Century Culture*, vol. 9, ed. Roseann Runte, University of Wisconsin Press, Madison, 1979, pp. 237–49.
Donovan, Arthur, ed., *The Chemical Revolution: Essays in Reinterpretation, Osiris*, second series, vol. 4, History of Science Society, Philadelphia, 1988; distributed by the University of Chicago Press.
Dowd, David Lloyd, *Pageant-Master of the Republic – Jacques-Louis David and the French Revolution*, University of Nebraska, Lincoln, 1948.
Doyle, William, *The Oxford History of the French Revolution*, Oxford University Press, Oxford, 1989.
Dujarric de la Rivière, R., *Lavoisier économiste*, Masson, Paris, 1949.
Durand, Yves, *Les fermiers généraux au XVIIIe siècle*, Presses Universitaires de France, Paris, 1971.
Duveen, Denis I., "Lavoisier and the French revolution," *Journal of Chemical Education*, part 1, 31 (1954), pp. 60–5; part 2, 34 (1957), pp. 502–3; part 3, 35 (1958), pp. 233–4; part 4, 35 (1958), pp. 470–1.
—— "Madame Lavoisier," *Chymia*, 4 (1953), pp. 13–29.
—— *Supplement to a Bibliography of the Works of Antoine Laurent Lavoisier 1743–1794*, Dawson, London, 1965.
Duveen, Denis I. and Klickstein, Herbert S., "Antoine Laurent Lavoisier's contributions to medicine and public health," *Bulletin of the History of Medicine*, 29 (1955), pp. 164–79.
—— "Benjamin Franklin (1706–1790) and Antoine Laurent Lavoisier (1743–1794)," *Annals of science*, 11 (1955), pp. 103–28, 271–308.
—— *A Bibliography of the Works of Antoine Laurent Lavoisier 1743–1794*, Dawson, London, 1954.
Ellery, Eloise, *Brissot de Warville: A Study in the History of the French Revolution*, Boston, 1915.
Eyles, V. A., "The evolution of a chemist – Sir James Hall, Bt., F.R.S., P.R.S.E., of Dunglass, Haddingtonshire, (1761–1832), and his relations with Joseph Black, Antoine Lavoisier, and other scientists of the period," *Annals of Science*, 19 (1963), pp. 153–83.
Fauré, Edgar, *La Disgrâce de Turgot*, Gallimard, Paris, 1961.
Fayet, Joseph, *La Révolution Française et la Science 1789–1795*, Marcel Rivière,

Paris, 1960.

Fichman, Martin, "French Stahlism and chemical studies of air, 1750–1770," *Ambix*, 18 (1971), pp. 94–122.

Fitzsimmons, Michael P., *The Parisian Order of Barristers and the French Revolution*, Harvard University Press, Cambridge, 1987.

Fox, Robert, "The rise and fall of Laplacian physics," *Historical Studies in the Physical Sciences*, 4 (1974), pp. 89–136.

Fox-Genovese, Elizabeth, *The Origins of Physiocracy*, Cornell University Press, Ithaca, 1976.

Frängsmyr, Tore, Heilbron, J. L., and Rider, Robin E., eds. *The Quantifying Spirit in the 18th Century*, University of California Press, Berkeley, 1990.

Fric, René, "Contribution à l'étude de l'évolution des idées de Lavoisier sur la nature de l'air et sur la calcination des métaux," *Archives Internationales d'Histoire des Sciences*, 12 (1959), pp. 136–68.

—— "Une letter inédite de Lavoisier à B. Franklin," *Bulletin historique de l'Auvergne, publié par l'Académie des Sciences, Belles-Lettres et Arts de Clermont-Ferrand*, 2nd series, no. 9 (1924), pp. 145–52.

Furet, François, "Terror," in *A Critical Dictionary of the French Revolution*, ed. François Furet and Mona Ozouf, trans. Arthur Goldhammer, Harvard University Press, Cambridge, 1989, pp. 137–50.

Gago, Ramón, "The new chemistry in Spain," in A. Donovan, ed., *Chemical Revolution*, pp. 169–92.

Gibbs, F. W., *Joseph Priestley*, Nelson, London, 1965.

Gillispie, Charles Coulston, "Carnot, Lazare Nicolas," in DSB, 3:70–9.

—— "The *Encyclopédie* and the Jacobin philosophy of science: a study in ideas and consequences," in *Critical Problems in the History of Science*, ed. Marshall Clagett, University of Wisconsin Press, Madison, 1959, pp. 255–89.

—— *The Montgolfier Brothers and the Invention of Aviation 1783–1784*, Princeton University Press, Princeton, 1983.

—— *Science and Polity in France at the End of the Old Regime*, Princeton University Press, Princeton, 1980.

Gingerich, Owen, "Lacaille, Nicolas-Louis de," in DSB, 7:542–5.

Golinski, Jan, *Science as Public Culture – Chemistry and Enlightenment in Britain, 1760–1820*, Cambridge University Press, Cambridge, 1992.

Gottschalk, Louis, *Jean Paul Marat: A Study in Radicalism*, 2nd edn, University of Chicago Press, Chicago, 1967.

Gough, J. B., "Lavoisier's early career in science: an examination of some new evidence," *British Journal for the History of Science*, 4 (1968), pp. 52–7.

—— "Lavoisier's memoirs on the nature of water and their place in the chemical revolution," *Ambix*, 30 (1983), pp. 89–106.

—— "The origins of Lavoisier's theory of the gaseous state," in H. Woolf, ed., *The Analytic Spirit*, pp. 15–39.

—— "Réaumur, René-Antoine Ferchault de," in DSB, 11:327–35.

—— "Some early references to revolutions in chemistry," *Ambix*, 29 (1982), pp.

106–9.

Goupil, Michelle, et al., eds, *Lavoisier et la révolution chimique*, SABIX: Paris, 1992.

Greenbaum, Louis S., "The humanitarianism of Antoine Laurent Lavoisier," *Studies on Voltaire and the Eighteenth Century*, 88 (1972), pp. 651–75.

Greene, John C., *American Science in the Age of Jefferson*, Iowa State University Press, Ames, 1984.

Gregory, Frederick, "Romantic Kantianism and the end of the Newtonian dream in chemistry," *Archives Internationales d'Histoire des Sciences*, 34 (1984), pp. 108–23.

Grimaux, Edouard, *Lavoisier 1743–94, d'après sa correspondance, ses manuscrits, ses papiers de famille et d'autres documents inédits*, 3rd edn, Alcan, Paris, 1899.

Guerlac, Henry, *Antoine-Laurent Lavoisier – Chemist and Revolutionary*, Scribner's, New York, 1975; first published, without illustrations, in *DSB*, 8 (1973).

—— "Chemistry as a branch of physics: Laplace's collaboration with Lavoisier," *Historical Studies in the Physical Sciences*, 7 (1976), pp. 193–276.

—— "Commentary on the papers of Charles Coulston Gillispie and L. Pearce Williams," in *Critical Problems in the History of Science*, ed. Marshall Clagett, University of Wisconsin Press, Madison, 1959, pp. 317–20.

—— *Essays and Papers in the History of Modern Science*, Johns Hopkins University Press, Baltimore, 1977.

—— *Lavoisier – The Crucial Year*, Cornell University Press, Ithaca, 1961.

Guerlac, Henry and Perrin, Carl, "A Chronology of Lavoisier's Career," typescript, n.d., n.p.; copy in Cornell University History of Science Collections.

Guyton de Morveau, L. B., *Digressions académiques*, Dijon, 1772.

—— *Élémens de chymie, théorique et pratique, rédigés dans un nouvel ordre, d'après les découvertes modernes, pour servir aux Cours publics de l'Académie de Dijon*, 3 vols., Dijon, 1777–8.

Guyton de Morveau, L. B., Lavoisier, A. L., Berthollet, C. L., and Fourcroy, A. F. de, *Méthode de Nomenclature chimique*, Cuchet, Paris, 1787.

Hahn, Roger, *The Anatomy of a Scientific Institution – The Paris Academy of Sciences, 1666–1803*, University of California Press, Berkeley, 1971.

Halleux, R., "La révolution Lavoisienne en Belgique," in Goupil, *Lavoisier*, pp. 295–311.

Hampson, N., *Will and Circumstance: Montesquieu, Rousseau and the French Revolution*, London, 1983.

Hankins, Thomas L., *Science and the Enlightenment*, Cambridge University Press, Cambridge, 1985.

Hanks, Leslie, *Buffon avant l'Histoire Naturelle*, Presses Universitaires de France, Paris, 1966.

Hannaway, Owen, *The Chemists and the Word*, Johns Hopkins University Press, Baltimore, 1975.

Heilbron, J. L., *Electricity in the 17th and 18th Centuries*, University of California

Press, Berkeley, 1979.

—— "Introductory essay," in T. Frängsmyr, *Quantifying Spirit*, pp. 1–23.

—— "The measure of enlightenment," in T. Frängsmyr, *Quantifying Spirit*, pp. 207–42.

—— "Nollet, Jean-Antoine," in DSB, 10:145–8.

Holmes, Frederic Lawrence, *Eighteenth-Century Chemistry as an Investigative Enterprise*, Office of History of Science and Technology, University of California, Berkeley, 1989.

—— *Lavoisier and the Chemistry of Life*, University of Wisconsin Press, Madison, 1985.

—— "Lavoisier's conceptual passage," in A. Donovan, ed., *Chemical Revolution*, pp. 82–92.

Home, R. W., "The notion of experimental physics in early eighteenth-century France," in *Change and Progress in Modern Science*, ed. J. C. Pitt, Reidel, Dordrect, 1985, pp. 107–31.

Hufbauer, Karl, *The Formation of the German Chemical Community (1720–1795)*, University of California Press, Berkeley, 1982.

Itard, Jean, "Clairaut, Alexis-Claude," in DSB, 3:281–6.

Jefferson, Thomas, *The Papers of Thomas Jefferson*, ed. Julian P. Boyd, 20 vols., Princeton University Press, Princeton, 1950–82.

Kaplan, Steven L., *Bread, Politics, and Political Economy in the Reign of Louis XV*, 2 vols., Nijhoff, The Hague, 1976.

Kersaint, Georges, *Antoine François de Fourcroy (1755–1809)*, CNRS and Muséum National d'Histoire Naturelle, Paris, 1966.

Kennedy, Emmet, *A Cultural History of the French Revolution*, Yale University Press, New Haven, 1989.

Kirwan, Richard, *An Essay on Phlogiston, and the Constitution of Acids*, new edn, J. Johnson, London, 1789.

Kremer, Richard, "Defending Lavoisier: the French Academy's prize competition of 1821," *History and Philosophy of the Life Sciences*, 8 (1986), pp. 41–65.

Kurzweil, Allen, "Laboratory of the soul," *Art and Antiques*, 7 (1990), pp. 90–5, 124–5.

Lacaille, Abbé de, *Leçons élémentaires de Mathématique*, 2nd edn, Paris, 1778.

Langins, Janis, "Hydrogen production for ballooning during the French revolution: an early example of chemical process development," *Annals of Science*, 40 (1983), pp. 531–58.

Lavoisier, Antoine Laurent, *Correspondance* (vol. 7 of *Oeuvres*), fascs 1–3 (1768–1783), ed. René Fric, Albin Michel, Paris, 1955–64; fasc. 4 (1784), ed. Michelle Goupil, Belin, Paris, 1986.

—— *Elements of Chemistry*, trans. Robert Kerr, Edinburgh, 1790; reprint, Dover, New York, 1965.

—— *Oeuvres de Lavoisier*, 6 vols., vols. 1–4, ed. J. B. Dumas, vols. 5–6, ed. Edouard Grimaux, Imprimerie Impériale, Paris, 1862–93; reprint, Johnson Reprint Corporation, New York, 1965.

Lavoisier, A. L., and Laplace, P. S., *Memoir on Heat*, trans. with introd. Henry Guerlac, Neale Watson, New York, 1982.

Le Grand, Homer E., "The 'conversion' of C. L. Berthollet to Lavoisier's chemistry," *Ambix*, 22 (1975), pp. 58–70.

Leicester, H. M., "The spread of the theory of Lavoisier in Russia," *Chymia*, 5 (1959), pp. 138–44.

Lenglen, M., *Lavoisier agronome*, Bulletin des Engrais, Paris, 1936.

Levere, Trevor H., "Balance and gasometre in Lavoisier's chemical revolution," in Goupil, *Lavoisier*, pp. 313–32.

Lundgren, Anders, "The changing role of numbers in 18th-century chemistry," in T. Frängsmyr, *Quantifying Spirit*, pp. 245–66.

—— "The new chemistry in Sweden," in A. Donovan, ed., *Chemical Revolution*, pp. 146–68.

McEvoy, John G., "Continuity and discontinuity in the chemical revolution," in A. Donovan, ed., *Chemical Revolution*, pp. 195–213.

McKie, Douglas, *Antoine Lavoisier*, Collier, New York, 1962; originally published by Constable, London, 1952.

Macquer, P. J., *Dictionnaire de chymie*, 2 vols., Paris, 1766.

Matthews, George T., *The Royal General Farms in Eighteenth-Century France*, Columbia University Press, New York, 1958.

Mauskopf, Seymour H., "Chemistry and Cannon: J.-L. Proust and gunpowder analysis," *Technology and Culture*, 31 (1990), pp. 398–426.

—— "Gunpowder and the chemical revolution," in A. Donovan, ed., *Chemical Revolution*, pp. 93–118.

Meldrum, Andrew Norman, "Lavoisier's early work in science 1763–1771," *Isis*, parts I and II, 19 (1933), pp. 330–63; part III, 20 (1934), pp. 396–425.

Metzger, Hélène, *Newton, Stahl, Boerhaave et la doctrine chimique*, Blanchard, Paris, 1930.

Morris, Gouverneur, *A Diary of the French Revolution*, ed. Beatrix Cary Davenport, 2 vols, Houghton Mifflin, Boston, 1939.

Morris, Robert J., "Lavoisier and the caloric theory," *British Journal for the History of Science*, 6 (1972), pp. 1–38.

Multhauf, Robert P., "The French crash program for saltpeter production, 1776–94," *Technology and Culture*, 12 (1971), pp. 163–81.

Nollet, Jean Antoine, *Leçons de physique expérimentale*, 6 vols., Paris, 1743–8, and later editions.

Nora, Pierre, "Republic," in *A Critical Dictionary of the French Revolution*, ed. François Furet and Mona Ozouf, trans. Arthur Goldhammer, Harvard University Press, Cambridge, 1989, pp. 792–805.

Outram, Dorinda, "The ordeal of vocation: the Paris Academy of Sciences and the Terror," *History of Science*, 2 (1983), pp. 251–73.

Partington, J. R., *A History of Chemistry*, vol. 3, Macmillan, London, 1962.

—— *A History of Greek Fire and Gunpowder*, Cambridge University Press, Cambridge, 1960.

Payen, Régis, *L'Evolution d'un monopole: l'industrie des poudres avant la loi du 13 Fructidor An V*, Editions Domat-Mont Christian, Paris, 1935.

Perrin, C. E., "The chemical revolution: shifts in guiding assumptions," in *Scrutinizing Science*, ed. Arthur Donovan, Larry Laudan and Rachel Laudan, Kluwer, Dordrecht, 1988, pp. 105–24; reprint, Johns Hopkins University Press, 1992.

—— "Document, text and myth: Lavoisier's crucial year revisited," *British Journal for the History of Science*, 22 (1989), pp. 3–25.

—— "Early opposition to the phlogiston theory: two anonymous attacks," *British Journal for the History of Science*, 5 (1970), pp. 128–44.

—— "Lavoisier, Monge and the synthesis of water," *British Journal for the History of Science*, 6 (1973), pp. 424–8.

—— "The Lavoisier-Bucquet collaboration: a conjecture," *Ambix*, 36 (1989), pp. 5–13.

—— "Lavoisier's table of the elements: a reappraisal," *Ambix*, 20 (1973), pp. 95–105.

—— "Lavoisier's thoughts on calcination and combustion," *Isis*, 77 (1986), pp. 647–66.

—— "Prelude to Lavoisier's theory of calcination. Some observations on *Mercurius Calcinatus per se*," *Ambix*, 16 (1969), pp. 140–51.

—— "A reluctant catalyst: Joseph Black and the Edinburgh reception of Lavoisier's chemistry," *Ambix*, 29 (1982), pp. 141–76.

—— "The triumph of the antiphlogistians," in H. Woolf, ed., *The Analytic Spirit*, pp. 40–63.

Pigeonneau, Henri, and de Foville, Alfred, *Procès-verbaux de l'administration de l'agriculture au contrôle général des finances, 1785–1787*, Paris, 1882.

Popkin, Jeremy, "Marat and the eighteenth-century science of violence," paper read at the Eighteenth-Century Seminar, State University of New York at Stony Brook, May, 1991.

Pouchet, Georges, *Les sciences pendant la terreur, d'aprés les documents du temps et les pièces des Archives nationales*, Société de l'Histoire de la Révolution Française, Paris, 1896.

Priestley, Joseph, *Experiments and Observations on Different Kinds of Air*, 3 vols., J. Johnson, London, 1774–7.

—— *A Scientific Autobiography of Joseph Priestley (1733–1804)*, ed. Robert E. Schofield, MIT Press, Cambridge, 1966.

Rappaport, Rhoda, "The early disputes between Lavoisier and Monnet," *British Journal for the History of Science*, 4 (1969), pp. 233–44.

—— "The geological atlas of Guettard, Lavoisier, and Monnet: conflicting views of the nature of geology," in *Toward a History of Geology*, ed. Cecil J. Schneer, MIT Press, Cambridge, 1969, pp. 272–87.

—— "Guettard, Jean-Étienne," in DSB, 5:577–9.

—— "Lavoisier's geologic activities," *Isis*, 58 (1967), pp. 375–84.

—— "Lavoisier's theory of the earth," *British Journal for the History of Science*, 6

(1973), pp. 247–60.

—— "The liberties of the Paris Academy of Sciences, 1716–1785," in H. Woolf, ed., *The Analytic Spirit*, pp. 225–53.

—— "G. F. Rouelle: an eighteenth-century chemist and teacher," *Chymia*, 6 (1960), pp. 68–101.

—— "Rouelle and Stahl – the phlogistic revolution in France," *Chymia*, 7 (1961), pp. 73–102.

Roberts, Lissa, "A word and the world: the significance of naming the calorimeter," *Isis*, 82 (1991), pp. 198–222.

Roche, Daniel, "Académies et politique au siècle des lumières: les enjeux pratiques de l'immortalité," in *The Political Culture of the Old Regime*, ed. Keith Michael Baker, Pergamon, Oxford, 1987, pp. 331–43.

Roger, Jacques, "Buffon, Georges-Louis Leclerc, comte de," in DSB, 2:576–82.

—— "Chimie et biologie: des 'molécules organiques' de Buffon à la 'physico-chimié' de Lamarck," *History and Philosophy of the Life Sciences*, 1 (1979), pp. 43–64.

—— "Diderot et Buffon en 1749," *Diderot Studies*, 4 (1963), pp. 221–36.

Rouelle, Guillaume François, "Traité de chymie de Rouelle," 2 vols. in 1, Paris, c.1762, Manuscript, Cornell University Library.

Sadoun-Goupil, Michelle, *Le chimiste Claude-Louis Berthollet, 1748–1822, sa vie, son oeuvre*, Vrin, Paris, 1977.

Saricks, Ambrose, *Pierre Samuel Du Pont de Nemours*, University of Kansas Press, Lawrence, 1965.

Schaffer, Simon, "Measuring virtue: eudiometry, enlightenment and pneumatic medicine," in *The Medical Enlightenment of the Eighteenth Century*, ed. Andrew Cunningham and Roger French, Cambridge University Press, Cambridge, 1990, pp. 281–318.

Schama, Simon, *Citizens*, Knopf, New York, 1989.

Scheler, Lucien, *Lavoisier*, Seghers, Paris, 1964.

—— "Lavoisier et la Régie des Poudres," *Revue d'Histoire des Sciences et de leurs Applications*, 26 (1973). pp. 194–222.

—— *Lavoisier et la Révolution française*, 2 vols, Hermann, Paris, 1960.

Schelle, G., and Grimaux, E., *Lavoisier – statistique agricole et projets de réformes*, Guillaumin, Paris, 1894.

Schofield, Robert E., "Priestley, Joseph," in DSB, 11:139–47.

Shapin, Steven, and Schaffer, Simon, *Leviathan and the Air-Pump*, Princeton University Press, Princeton, 1985.

Shennan, J. H., *The Parlement of Paris*, Eyre and Spottiswoode, London, 1968.

Siegfried, Robert, "The chemical revolution in the history of chemistry," in A. Donovan, ed., *Chemical Revolution*, pp. 34–50.

—— "Lavoisier's table of simple substances: its origins and interpretation," *Ambix*, 29 (1982), pp. 29–48.

—— "Lavoisier's view of the gaseous state and its early application to pneumatic chemistry," *Isis*, 63 (1972), pp. 59–78.

Smeaton, W. A., "L'Avant-Coureur. The journal in which some of Lavoisier's earliest research was reported," *Annals of Science*, 13 (1957), pp. 219–34.

—— "The contributions of P. J. Macquer, T. O. Bergman and L. B. Guyton de Morveau to the reform of chemical nomenclature," *Annals of Science*, 10 (1954), pp. 87–106.

—— "The early years of the Lycée and the Lycée des Arts. A chapter in the lives of A. L. Lavoisier and A. F. de Fourcroy," *Annals of Science*, 11 (1955), pp. 257–67.

—— *Fourcroy, Chemist and Revolutionary, 1755–1809*, Heffer, Cambridge, 1962.

—— "Geoffroy, Étienne François," in DSB, 6:352–4.

—— "Guyton de Morveau, Louis Bernard," in DSB, 5:600–4.

—— "Guyton de Morveau and the phlogiston theory," in *Mélanges Alexandre Koyré*, ed. I. B. Cohen and René Taton, 2 vols, Hermann, Paris, 1964, pp. 523–40.

—— "Lavoisier's membership of the Assembly of Representatives of the Commune of Paris," *Annals of Science*, 13 (1957), pp. 235–48.

—— "Lavoisier's membership of the Société Royale d'Agriculture and the Comité d'Agriculture," *Annals of Science*, 12 (1956), pp. 267–77.

—— "Madame Lavoisier, P. S. and E. I. Du Pont de Nemours and the publication of Lavoisier's 'Mémoires de chimie'," *Ambix*, 36 (1989), pp. 22–30.

—— "Monsieur and Madame Lavoisier in 1789: the chemical revolution and the French revolution," *Ambix*, 36 (1989), pp. 1–4.

—— "Some large burning lenses and their use by 18th century French and British chemists," *Annals of Science*, 44 (1987), pp. 265–76.

—— "Venel, Gabriel François," in DSB, 13:602–4.

Snelders, H. A. M., "The new chemistry in the Netherlands," in A. Donovan, ed., *Chemical Revolution*, pp. 121–45.

Stone, Bailey, *The French Parlements and the Crisis of the Old Regime*, University of North Carolina Press, Chapel Hill, 1986.

Storrs, F. C., "Lavoisier's technical reports, 1768–1794," *Annals of Science*, 22 (1966), pp. 251–75; 24 (1968), pp. 179–97.

Sutton, Geoffrey, "Electrical medicine and mesmerism," *Isis*, 72 (1981), pp. 375–92.

Thackray, Arnold, *Atoms and Powers*, Harvard University Press, Cambridge, 1970.

Torlais, Jean, *L'Abbé Nollet, 1700–1770, un physicien au siècle des lumières*, Sipuco, Paris, 1954.

—— "La physique expérimentale," in *Enseignement et Diffusion des sciences en France au XVIIIe Siècle*, ed. René Taton, Hermann, Paris, 1964, pp. 619–45.

Turgot, A. R. J., "Expansibilité," in *Encyclopédie, ou Dictionnaire raisoné des sciences, des arts et des métiers, par une société de gens de lettres*, ed. Denis Diderot and Jean d'Alembert, vol. 6, Paris, 1756, pp. 274–85.

Venel, Gabriel François, "Chymie," in *Encyclopédie, ou Dictionnaire raisoné des sciences, des arts et des métiers, par une société de gens de lettres*, ed. Denis Diderot and Jean d'Alembert, vol. 3, Paris, 1753, pp. 408–47.

Woolf, Harry, ed., *The Analytic Spirit*, Cornell University Press, Ithaca, 1981.

Index

Page numbers in italics refer to illustrations or their captions.

Milton Keynes UK
Ingram Content Group UK Ltd.
UKHW041523181024
449640UK00009B/175

9 780521 566728